Hogs Are Up

Hogs Are Up

Stories of the Land,
with Digressions

WES JACKSON

Foreword by Robert Jensen

UNIVERSITY PRESS OF KANSAS

For Ken Warren

Published by the University Press of Kansas (Lawrence, Kansas 66045),
which was organized by the Kansas Board of Regents and is operated
and funded by Emporia State University, Fort Hays State University,
Kansas State University, Pittsburg State University, the University of Kansas,
and Wichita State University.

Library of Congress Cataloging-in-Publication Data
Names: Jackson, Wes, 1936– author. | Jensen, Robert, 1958– foreword.
Title: Hogs are up : stories of the land, with digressions / by Wes
Jackson, foreword by Robert Jensen.
Description: Lawrence, KS : University Press of Kansas, [2021]
Identifiers: LCCN 2020036223
ISBN 9780700630592 (cloth)
ISBN 9780700630608 (ebook)
Subjects: LCSH: Jackson, Wes, 1936– | Human ecology. |
Radicals—Kansas—Biography. | Deep ecology.
Classification: LCC GF41 .J324 2021 | DDC 304.2—dc23
LC record available at https://lccn.loc.gov/2020036223.
British Library Cataloguing-in-Publication Data is available.
Printed in the United States of America
10 9 8 7 6 5 4 3 2 1

The paper used in this publication is acid free and meets the minimum
requirements of the American National Standard for Permanence of Paper
for Printed Library Materials Z39.48-1992.

Contents

Foreword: "Tell about It"

Robert Jensen

The best summary of Wes Jackson's method for inquiring into the world that I have ever found comes from Mary Oliver, who in her poem "Sometimes" offered "instructions for living a life":

> Pay attention.
> Be astonished.
> Tell about it.[1]

Whether wrangling cows on a patch of Kansas pasture or wrestling with scientific accounts of the origins of life on Earth, Jackson pays attention, is constantly astonished by the world around him, and can't wait to tell someone about it.

Pay attention to everything around and inside you. Pay attention to plants and other creatures, both out in "nature" and on city streets, because life is everywhere. Pay attention to everyone you meet, even in routine interactions, and never pass up a chance to ask someone what they hope to accomplish in the world. Pay attention to your own reactions to people and places, learning from what attracts and repels you.

Opportunities for astonishment are everywhere if you are paying attention, not only in the extraordinary but also in the everyday. It certainly is astonishing how mountains arose from the crashing of tectonic plates, how the great civilizations of the ancient world rose to power, fell, and disappeared. But just as astonishing is the way that plants can grow in the small amount of soil in the cracks of city side-walk and the way that love endures despite human cruelty.

Jackson pays attention and is constantly astonished, qualities that he shares with many. What makes him distinctive is how he tells about it. Jackson has written scholarly articles, books, and grant proposals followed up by reports on the projects that those grants funded. But his natural vehicle for telling about it is a story. Mention almost any topic, and Jackson will say, "Well, yes, and I've got a story about that," and indeed he does. A long life of paying attention and being astonished, combined with an unusually sharp memory for details from years long past, has left Jackson with a seemingly endless supply of stories that entertain, enrich, and educate. Whether in a formal lecture or casual conversation, Jackson cannot stop himself from telling stories.

Everyone can, and does, tell stories, of course—it is part of what makes us human. Some people are verbose and some quiet, but we are always telling stories in our heads, and one of the great joys of life is sharing those stories. Even the shyest of people cannot completely repress the desire to tell a story. The literature scholar Jonathan Gottschall—who describes us as *Homo fictus,* "the great ape with the storytelling mind"—puts it this way: "We are, as a species, addicted to story. Even when the body goes to sleep, the mind stays up all night, telling itself stories."[2]

Ask Jackson how he slept the previous night, and a common answer these days is "Not well. My head was buzzing all night." As Jackson gets older, it is as if he wants to make sure no story goes untold.

In some ways, our storytelling obsession should not be surprising, given our evolutionary history. The paleoanthropologist Ian Tattersall puts it this way: "For all the infinite cultural variety that has marked the long road of human experience, if there is one single thing that above all else unites all human beings today, it is our symbolic capacity: our common ability to organize the world around us into a vocabulary of mental representations that we can recombine in our minds, in an endless variety of new ways."[3]

In other words: We are a storytelling animal. The question is not whether we will tell them but what kind of stories we will tell, with what spirit, and toward what end. Jackson's most memorable are stories of his astonishment, which is a way of saying stories of his intellectual life.

In Defense of Intellectual Life

Jackson is, in the performative sense, a great storyteller, as those who have heard him speak or shared his company will confirm. That skill is reflected in the written stories in this volume, but my interest is less in performance than in ideas. What makes Jackson's storytelling important at this moment is the way he puts stories to use in building and sharing his intellectual life.

First, a word about "intellectual," a term that may trigger an instinct to snap this book shut. That's likely because there are people who use that word arrogantly, to assert superiority (think of the most annoying college professor you have ever met or anyone with a PhD who demands to be called "Dr."). But everyone can embrace the joy that comes with the life of the mind.

In this context, we'll define intellectual work simply as a systematic effort to (1) pose interesting questions about the world around us, (2) collect information relevant to those questions, (3) analyze that information to discern patterns that help deepen our understanding of how the world works, and (4) use that understanding to make judgments about how to try to shape a better world.

In more concise form, intellectual work is what, how, and why. What is the world? How does it work? Why does it work that way? No one needs a PhD to ponder those questions.

Understood this way, the issue is not "Who is an intellectual?" but rather "How well are we all doing the intellectual components of our work?" For me, Jackson has been an intellectual model, not because he's always right about everything but because of the joyful spirit behind his inquiry and his openness to being challenged. He has sharpened his intellectual abilities in traditional academic settings, earning a BA in biology, an MA in botany, and a PhD in genetics. His intellectual work has led to awards—the MacArthur Foundation "genius grant," the Right Livelihood Award, one of the Smithsonian's "35 Who Made a Difference," and on and on. But he knows that folks who have never set foot on a university campus or never won an award often do first-rate intellectual work, and he has learned from them all.

On one June day that I was visiting Salina, here's how I saw Jackson's intellectual life in action, in very different settings.

Jackson has a few head of cattle on his land, and that day I watched him and others try to load two of those cows into a trailer to take to market. That project required understanding the psychology of cows in general, along with knowing the particular traits of not only those two cows but also the others who weren't heading to sale. That knowledge was relevant in designing a route from the pasture to the trailer and using the inducement of feed and metal gates to manage the flow, all while constantly assessing the slipperiness of the rain-soaked ground to avoid losing footing.

Later that same day, Jackson was in his office, cocooned by bookshelves that hold a half century of eclectic reading, hashing over one of the most basic questions not only in biology but in all of human life—how did life come to be? What story can we tell about "the journey from minerals to cells," one of Jackson's favorite phrases to capture the question? Pecking away at his manual Royal typewriter, Jackson was working on an essay to describe how elusive the answer is, and perhaps always will be.

No one would hesitate to call the latter effort "intellectual work." But anyone who went after those cows without recognizing the intellectual work—the task of understanding all those factors and their interaction—required to load those cows would likely end up impaled on a longhorn.

Too often, people associate intellectual work with "just thinking," separate from the work of feeding and sheltering ourselves through daily activity. Jackson understands that all of this activity requires intellectual work, which can be done well or poorly. Some of his intellectual role models had no merit badges from universities and no recognition from the formal culture. Other important teachers were paid to do intellectual work in the lab and the classroom. All are part of a rich intellectual life, as the stories in this book make clear.

Just Visiting

It's all well and good to try to take the snobbishness out of the term "intellectual," but there will always be people who use the term to suggest that they outrank others in the thinking department. So in some settings it's probably wise to drop the term from conversation. Instead of presenting the stories in this book as intellectual engagement, we can think of them as "just visiting" with Jackson.

In the world in which Jackson grew up—before television, before cable, before the Internet, before streaming services—people did a lot more visiting. There were books and a radio in Jackson's childhood home, but much of the storytelling in that era was locally produced in face-to-face interaction, not piped in from professionals. The term "visiting" captures that culture.

To go visiting was more than simply dropping by the home of a neighbor or relative. To ask, "Did you have a good visit?" was an inquiry into the nature of the conversation. A good visit meant that people were engaged with each other—the conversation was the center of the time together. There might have been food and beverages served when company dropped by, but a good visit was dependent not on the quality of the snacks but rather on the quality of the stories told.

Before there were telephones in homes, folks couldn't call ahead to arrange a visit. In the world Jackson grew up in, even with phones, stopping by for a visit without making a plan ahead of time was not unusual. Visiting was part of community life. Jackson remembers his mother saying one Sunday afternoon, "Let's go for a drive," because she had not had time to clean the house, and she didn't want to entertain people when the house didn't meet her standards. No one was scheduled to come by, Jackson said, but you never knew when people might drop by to visit.

When stopping by for a visit, one might have a particular subject or question in mind, but the storytelling in most visits was free-flowing. An observation about the heavy rain that week might lead to a story about a memorable flood, which might spark a memory of a family that had lost their home that year, which might remind someone that the eldest daughter in that family had gone away to college in Lawrence, and didn't she end up marrying that boy who lived next door to your grandmother, who . . . and so it goes. Some might find the nonlinear nature of such conversation frustrating, but for Jackson much of the joy in storytelling comes from the digressions.

Hogs Are Up: A Digression about Digressions

If you were wondering when the title of this book would be explained, we're there.

Every family has words and phrases that become a kind of shorthand for a longer thought. Such was the case with "Hogs are up." In some conversation long ago, Jackson said that his mother interrupted the flow by announcing, without any introduction, "Hogs are up." She had just heard commodity prices on the radio and felt that people needed to know it.

From that point forward, when anyone in the family had something important to say that was out of the flow of the conversation going on, they may announce, "Hogs are up" and then state their business. Jackson remembered this when we were editing the manuscript, and I came to a place where I thought a transition was needed to explain the connection of two ideas. Without giving it a thought, Jackson said, "Hogs are up." I was confused by the apparent non sequitur but laughed when he told the story in far richer and funnier fashion than I just summarized.

So apparently his penchant for digressions has a genetic component. In many a lecture Jackson will say, "I feel a digression coming on." Audiences figure out quickly that that means it is time to pay close attention—his digressions are often the best part.

For Jackson, digressions are not a problem to be suppressed but part of an educational philosophy. When The Land Institute (TLI) began its work as an alternative school, people would often ask why he felt the need for something different. The former high school teacher and college professor said he worried that in traditional education "students seemed to be given more to minimal compliance than spontaneous elaboration." At TLI there were no exams or traditional assignments, just an invitation to collaborate on common problems through classroom work and time in the fields and shop. That strategy eliminated the temptation to shoot for minimal compliance and expanded the space for spontaneous elaboration, which does not always unfold in linear fashion.

In the stories in this book, expect digressions and give them your full attention. Jackson worked hard to make sure he did not leave out any of those digressions. While he feels free to digress in conversation and public talks, Jackson worried that in writing, those digressions would annoy readers. As he was writing, he frequently asked, "Have I taken too many off-ramps in this one?" My answer was always no. Intellectual life—even the intellectual life of the most rigorous scientist who meticulously follows an experimental protocol—is full of digressions, loops, dead-end ideas, and unexplainable insights. Much of that messiness gets cleaned up in articles written to report the results. In this book Jackson's goal is not just to present conclusions but to demonstrate the wonderfully anarchic working of the human mind.

The Assignment

I am a retired teacher, with no authority to give anyone an assignment. But I can't help myself. While these stories are amusing, readers should not settle for being amused. Here's your assignment: Pay attention to the stories, be astonished, and tell others about it.

The assignment is not as simple as it sounds. Sometimes "the moral of the story" may seem self-evident. "The Boy Who Cried Wolf" teaches us the consequences of sounding false alarms. "The Little Red Hen" reminds us of the perils of not doing our fair share of the work. But Jackson's stories are not constructed as moral fables for children. Running throughout this volume are the complexities, conundrums, and contradictions of being human while trying to understand the natural world and fellow humans who are a part of that world. In making environmental policy for the planet and in making a living for ourselves, we struggle to figure out what kind of relationship to the larger living world is appropriate. In the great questions of philosophy and the everyday decisions we have to make about how to treat people around us, we struggle to figure out what kind of creatures we are. That's intellectual work.

At the end of the same day that Jackson was herding cattle and ideas, we drove past the pasture so he could check on his cows. Separating out those two animals and sending them off to market weighed on Jackson, and he wanted to see how the remaining three cows were doing. He couldn't know exactly what was on those cows' minds, but he has affection for them and can imagine them missing their friends.

Once home, as we walked up the driveway, he turned and said, out of the blue, "Here's something I've been thinking about. How is it that I can look at some people, find no fault with them, see that they are decent people who share my values, and still not like them?" Jackson wasn't sure why that was on his mind just then, but it struck him as a puzzle. It's easy to know why we don't like people we think are mean, hateful, or unprincipled. But when someone seems to be "our kind of people," why might we still not like a person?

Struggling to understand cows and people, engaging our heart and mind, is intellectual work. And Jackson has a story about all those struggles. But do not expect that you will extract easy answers from these stories. Instead—if you pay attention and are open to being astonished—you will find yourself wanting to tell others about what Jackson's stories made you think. You will be doing intellectual work. And it will be fun.

Jackson comes from a farm family in which one measure of virtue was your ability to hoe weeds efficiently. He is a scientist who evaluates an idea based on the data from an experiment. He has built houses and barns and has great affection for the tools he used to build them. Jackson is fully engaged in hands-on fashion with the material world around him. At the end of the day, he likes to tot up the tangible accomplishments of that day.

But in the end, Jackson cares most deeply not about what has been accomplished today but about the multiple, cascading ecological crises that jeopardize tomorrow. Before he dies, he wants to see modern consumer culture change course toward a more sustainable and just human future. Toward that end, he comes back over and over to one question: "How do people change?" How will people come to see the need for a new relationship to each other and the larger living world?

In that endeavor, Jackson recognizes that the latest scholarly study and government report are not what motivate people to change. We have to learn to see the world in a different way. In the end, Jackson recognizes the truth at the heart of the poet Muriel Rukeyser's simple statement: "The universe is made of stories, not of atoms."[4]

Welcome to the world made of Wes Jackson's stories.

Where to Begin and End?

There is reason why writers seem to save expressions of gratitude until the end or near the end of a writing. They have to ponder who and what helped them approach or cross the finish line. A fear of the writer is to leave someone out of the acknowledgements, especially if the list is long. Indeed, how far back should one go to express his or her appreciation of places and times, as well as people? There is always a history of places and times, and much of the help is more than human.

Most of the people who are central in the stories of this book have long since died. Even so, we dare not try to escape from or fail to appreciate their influence. Any praise the dead mentioned in this volume deserve is for us to ponder, appreciate, and learn by.

But now, dear reader, you may be saying something like, "Come on, Wes, who alive helped you with this book?" Okay, that's a practical question. The answer begins with you, the reader. As I wrote I was hoping that the stories will be regarded as interesting, but that alone is not enough. I wanted the lives of the people represented in these pages to have meaning to you in some practical sense, as a kind of arrow pointing the way toward a cultural handing-down. Most of the people represented in this volume lived in a culture, probably without knowing it, that used energy and material resources "according to the holy dictate of spare temperance," to borrow a phrase from John Milton. I hope this all has relevance for us as we struggle to down power. We do have to learn to live within limits of energy and materials. The dead who worked in that time had lives which deserve some kind of credit. I was imagining readers who will appreciate those who, with dignity, did with less.

Now for those very-much-alive who assisted me with this volume. Robert Jensen put me up to writing this book of stories. I am beyond grateful for his insistence, his persistence, and his help. Bryan Thompson, my assistant, moved my non-electric Royal typewriter pages to the computer and offered necessary corrections and suggestions, for which I am appreciative.

I also owe a debt of gratitude to the people at the University Press of Kansas, especially Bethany Mowry, who was my editor. But Bethany had help, too, and that included people getting the pictures properly spaced, and the printing, and the publicity, and design: Kelly Chrisman Jacques, Mike Kehoe, Karl Janssen, Derek Helms, Erica Nicholson, and Jane Raese. Well, you can imagine the reality. I'm thankful for all at the Press.

Finally, I thank Joan, my wife, for her suggestions, all of which improved the text. Especially, I thank her for the photograph on the cover. Taken looking toward the river, it displays in part why "I sometimes wonder if the loss of Eden was a bargain," as I mention in the story, "Thoughts on the Natural History of Eden."

Introduction

----

*Getting the story right can be complicated.
Memory is unreliable. One person's perspective is
just that, the world as seen from one person's point
of view. A story can be accurate, even true, but still
be missing a key element that changes everything.
Sometimes a story is true, but then another truth
shows up, and the story is never the same after that.*

Year of Decision 1976: The Rest of the Story

We had been back in our house and home on twenty-eight acres just outside Salina for two years, property we had bought and built on when I was teaching at Kansas Wesleyan University in the late 1960s. A faculty position at California State University, Sacramento, which hired me to start an environmental studies department, took the family and me west in 1971, but we held on to the Kansas place. When I took a one-year leave, we went back to Kansas. The one year stretched into two years, and then came the call from Sacramento that I expected but dreaded. I was told unequivocally that I had to either return to my position in Sacramento or resign. It was as simple as that. From a professional point of view, in Sacramento I had a good job with tenure in the academic world. And so the choice: security in California, a decent salary, and health benefits. Security, money, and health benefits were not here in Kansas.

Sacramento, the capital city, had the clear water of the American River, which came right by the campus out of the western slope of the Sierra Nevada. We had several friends and lots of intellectual engagement, and if we wanted to do anything politically, we were in the capital of a large state where the levers of power for both the state and federal offices were within minutes. The Sierra Nevada loomed to the east, the Central Valley of California with its variety of agricultural crops lay to the south, and the Pacific Ocean with its public beaches was a short trip west. The University of California, Davis was less than a half hour away. We could go to San Francisco, the redwoods on the coast, the Central Valley, or Lake Tahoe and hike in the mountains along a trail toward the peaks. Lots going on in California for our family.

When we left for that leave to Kansas, we said that we planned to return to California after one year. But if that was the plan, why did we sell the house in Sacramento that bordered the American River, with its hiking/biking trail along the levee that bordered our backyard?

What could have been on our minds to want to return to Salina in the first place, even if for only a year? The next large city west of Salina was Denver, 435 miles away. The next larger-than-Salina town to the east was Topeka, some 110 miles away. Kansas City was nearly three hours away. No mountains, no ocean. No wide variety of fruits and vegetables. No Central Valley. No UC Davis close by, no UC Berkeley on East Bay, no Stanford on the peninsula.

What Salina had to offer was twenty-eight acres with a flat-roofed 1,300-square-foot house, partly finished, on the bank of the Smoky Hill River. The average rainfall was twenty-eight inches. Not bad, but it had been in the Dust Bowl corridor in the 1930s. Salina had also been a primary nuclear target for the USSR, given that Schilling Air Force Base had until recently had B-47 bombers loaded with nuclear weapons ready every night to head north to Russia, and there were a few intercontinental ballistic missile silos nearby. The base had recently been shut down, but the missile silos remained. The shutdown led to a population decline in Salina and a scramble to keep the economy going. For God's sake, what were we going back to?

We had family in and around Topeka. My mother was still alive, and from time to time we would be able to see her, my siblings, and their families, many of whom had traveled to visit us in California, where we would take them to see the redwoods, Chinatown, Fisherman's Wharf, ride the cable cars, go up Lombard Street. Why was being so close to home an excuse to stay?

As parents, we worried about the future and about how we might prepare ourselves and our children, developing the necessary skills for an uncertain future, which is to say, having the know-how to get along during periods of scarcity. We wanted to see what it would be like to live closer to subsistence level. We had the milk cow, the butcher hog, and chickens for meat and eggs, and my then wife, Dana, was a good gardener and became a good beekeeper. We put the kids to work, and all three learned skills without knowing it, the lessons built in to everyday chores. We had two years of this marginal life behind us—lumber buzzed up by a local sawmill and a shop made mostly from scrap lumber—and we somehow managed, and we were young.

Sacramento called: "We need to know. Return or resign." What to do? We had no money. I would do a welding job here and there. Dana did some substitute teaching. As we pondered our options, I thought back to the many nights in Sacramento that I had stared at the ceiling, thinking about the ideal school for college-aged students. I had read up on the Deep Springs College in California and Berea College in Kentucky, which combined intellectual rigor with hands-on work. I wondered if anything like that could sprout in Kansas. But with no money and little security, starting an alternative school wouldn't be simple.

Enter Kansas State Senator John Simpson, an attorney who lived in Salina. He and his then wife, Diane, invited us to a weenie roast in their yard. As John and I sat roasting our wieners and then marshmallows, he inquired about what we intended to do. I told him about the choice facing me and that I had been thinking of the ideal school. He quickly replied, "If you want to start a school, I'll help you." John offered to pay half the tuition for any student. Now we had an option: Stay and start a school. This required a family discussion, which included talk of the sort of school I imagined, but still there was the insecurity.

With John Simpson's offer on the table, the five of us—I, Dana, and our children, Laura, Scott and Sara—assembled in the house to discuss the pros and cons of staying. It seemed to me that it would be too risky to stay. If it didn't work out, well, the options would be greatly reduced. And so I said, "We'd better go back." At that point, Laura, age fifteen, burst into tears and said, "I thought you said that we are not called to success but to obedience to our vision." Hooboy! I resigned, and we stayed.

Countless times over the years I have been asked what led to the decision to leave California and start The Land Institute, and countless times I have told that story. There would be the sober acknowledgment that it was the wisdom of youth with high ideals that brought The Land into existence. "My, my, think of that" and "Out of the mouths of babes," "a child will lead us," and so on. For more than forty-three years, I told something close to this story. But as the old commentator Paul Harvey here in the Midwest would say every afternoon on the radio after the ad, "Now, the rest of the story."

On June 15, 2019, my wife, Joan, and I are in Cedar Falls, Iowa, in the home of Laura, her husband, Kamyar, and their daughter Ada. Nettie, their oldest at twenty-three and out of college, has come from Minneapolis. We are there to celebrate two birthdays, Kamyar's sixtieth two days later and my eighty-third that day. It is also Kamyar's fortieth year since leaving Iran. Some twenty-five or thirty friends and guests are seated here and there around the room, on the sofa, at tables, on the steps in groups of two, three, four, five. Someone asks me the question. I motion toward Laura with the back of my hand and say she's the reason. Then I tell the story once more. Once again the hearers seem amazed, offering thoughtful nods and such. There are looks toward Laura, but within seconds she describes a deeper truth that jump-started the past forty-three years of efforts at The Land.

"Well, there is more to that story," she says. "We had moved seven times, and I had gone to seven different schools. I didn't want to move again, and I knew that if I threw one of Dad's lines back at him, I wouldn't have to change schools again."

I was astounded. I had lived with a good story, my incomplete story of almost mythical proportions. It was the first time I had ever told the story with Laura present.

Well, Laura got what she wanted back then. The Land Institute came into existence because a fifteen-year-old girl did not want to change schools again.

There are stories, and there is the story.

Down from the College

I don't imagine that I was a planned child, being the last of six spread over twenty-two years from the same parents. There is reason to believe they wanted to stop at three, but that would have been in 1919. Well, you know how it is. It just kept happening. The next in line before me is eight years older. My father would turn fifty the year I was born and my mother forty-two. So I was something of a tagalong, arriving midyear in 1936. Mention the 1930s and it brings to mind both the Depression and the Dust Bowl for Kansas. Entry into World War II was five and a half years away. I was told that I was born on the farm, on the kitchen table, with the assistance of a twenty-year-old sister on her way to becoming a nurse.

In my early memories, men would come to our farm to give advice and help monitor a five-year experiment involving twenty-some crops. That required considerable data taking, mostly on the part of my dad but I suspect with the help of my older brothers. All of these professionals were, as family members put it, "down from the college," which meant Kansas State College, now University. These "down from the college" scientists were kindly men, respectful in every way. Beyond those visiting scientists, Kansas State was "our college." My older brother started school there in 1937 but came back home to farm after one semester. Since then several of my nieces, nephews, and cousins as well as a daughter and two grandchildren have been K-State students. My mother had a cousin who was president there for six years.

Beyond these personal engagements, there is another reason to respect the land-grant college along with the experiment station and extension service. These three together represent one of the greatest initiatives dedicated to the democratization of knowledge, given that the purpose was to include agriculture and the mechanical arts. In the mid-1960s I earned my PhD in the genetics department of

My parents, Howard and Nettie (Stover) Jackson, married December 31, 1912. After the wedding, they took a horse and buggy to their new home.

The family home in 1939. This house was designed by my mother and built by my father and brothers, with help from Uncle Roy.

another land-grant institution, North Carolina State. (The term "land grant" comes from the 1862 Morrill Act, which granted public lands for such a college in each state. The story of democratization is complicated, of course, by the dispossession of indigenous people behind the project.[1])

With all of these family connections, from birth through graduate school, I had reason to be a lover of the land-grant system. But somewhere during that journey, I began a quarrel with the entire institutional structure as I realized that it had become mostly a cheerleader for industrial agriculture. It's a lover's quarrel, and it endures even as The Land Institute—after years of being treated as a marginal enterprise by those folks—is today increasing its engagement with such universities.

But I spend less time arguing about the institutional structure now, as I have come to realize that most of the people are products of the system and unable to bring about the necessary fundamental change. If they ever did speak out about the destructive consequences of industrial agriculture, they long ago quit. Export policy, the commodity groups—in short, corporate power—will have their way. My daughter Laura identified this situation as a form of Stockholm syndrome (when hostages develop a connection to, and alliance with, their captors) for certain faculty who come to endorse the powers that have captured their university.

Countless painful moments of engagement have accumulated over the years. As I have gone to and fro on my ecological agriculture errands during the past nearly half century, the arguments have varied in their intensity. Early on I argued mostly with men my age or older. And now for something of the story behind how those inevitable disagreements arose during my forays into the ag schools of those land-grant universities, the story of how I went a different way.

Within a year after the start-up of The Land Institute, in the spring of 1977, I had two experiences within a short period of time. I had read the General Accounting Office (GAO) report on the effectiveness of the Soil Conservation Service. It looked to me, in reading that report, as if soil erosion was about as bad in the seventies as when the service started back in the thirties, when it was initially called the Soil Erosion Service. I wondered how this could be, given the thousands of miles of terraces, grass waterways, and such that the service had helped put in place. Shortly after that I took my interns to what came to be called the Konza Prairie near Manhattan, which is the home of Kansas State University. Lloyd Hulbert—a K-State biology professor, ecologist, internationally known expert on tallgrass prairies, and friend—led us on a field trip, describing that grassland ecosystem and mentioning the role of fire and grazing. It was a wonderful trip for all.

Back home, with the GAO study on my mind, I thought about how that prairie had no discernible soil erosion, no chemical contamination of the land or water, and no fossil-fuel dependency. The contrast between nature's prairie and human-designed agriculture was clear. Our grain crops, which grow on around two-thirds to three-fourths of our agricultural acreage and provide a similar share of our calories, are annual monocultures. The prairie features perennial species in mixtures, what we call polyculture.

There was a third element in this thought process. In the back of my mind was a single sentence from my graduate days in Raleigh, North Carolina, something my major professor had told me. Late one night, Ben W. Smith walked into my lab while I was at my microscope and said, "We need wilderness as a standard against which to judge our agricultural practices." He turned around and left without elaborating, but that idea stuck with me.

The GAO report could have been just another document on one of my countless unorganized piles, saved to study in more detail when I had the time. Luckily, it was enough on my mind that after the trip to the Konza Prairie, for whatever reason, I drew a branching diagram on the back of a brown paper grocery sack, starting with four paired contrasts: polyculture versus monoculture, perennial versus annual, woody versus herbaceous, and fruit/seed versus vegetative. I was looking for the combinations useful to humans. Four times four yields sixteen combinations. Four combinations were nonsensical (for example, woody annuals don't exist). Of the twelve remaining real possibilities, eleven of the blanks had been filled with plants useful for direct human purposes. But one combination, a polyculture of herbaceous perennials devoted to fruit/seed, was blank.

Why this was so, I could not get off my mind. I was mindful that nearly all of nature's land-based ecosystems featured perennial mixtures and knew something of the history of earth abuse due to agriculture, so it became clear that our failure to fill that blank stood behind the ten-thousand-year-old problem *of* agriculture (as opposed to the more familiar focus on specific problems *in* agriculture). The

Author with Ben W. Smith in December 1966, on the day I finished my final oral exam at North Carolina State.

same year I wrote a paper for our *Land Report* titled "The Search for a Permanent Agriculture." I later changed the title, replacing "permanent" with "sustainable," and published it in the Friends of the Earth journal called *Not Man Apart.* In 1980, Friends of the Earth published my book *New Roots for Agriculture*, an expansion of the idea and the argument supporting the necessity and possibility of a sustainable agriculture as I perceived it.

Well, the word was out, at least in the alternatives network, which was small enough at that time that we mostly all knew each other. The nonprofit organizations had little to no objection to my story. I was invited to give talks and seminars here and there at a few colleges, including some of the land-grant universities. Even though my degree from North Carolina State made me, in some sense, "one of them," I had to think carefully about how to present myself and my ideas, given the dominance of Big Ag at those schools.

Back in the seventies and eighties, there were many ag professors my age or older, born before or during the Great Depression, who had experienced the Dust Bowl years on a farm. I quickly learned that in a college of agriculture, professors speaking on a panel often made sure to establish their rural roots. If they had been born and/or raised on a farm, they said so. If they had gone to a one-room or two-room country school, so much the better. If there was no electricity on the farm when they were born and if the family had an outhouse, they would throw that in. If they worked your way through college washing dishes or went to a college of agriculture and milked cows at the state college dairy to cover tuition and living costs, they would by all means mention that.

Though I learned how important it was to establish one's humble rural origins—once again, I was one of them in this regard—that didn't help me all that much either. It turned out that most of those professors went on to explain that leaving those humble beginnings behind was a good thing, and there was no need to challenge the "progress" we had made in agriculture. We certainly didn't need any harebrained scheme that would take fifty years, maybe even a hundred, as I had predicted for perennial grains in my 1970s writings. The dominant view was that such a radical shift in agriculture was either not possible or not necessary.

In the last quarter of the last century, the Green Revolution was underway, big time. So were Big Ag, chemicals, and high-tech gadgets, which were dominating agriculture. Turns out that all my land-grant connections and rural-roots cultural capital didn't get me very far. I was seen by many as more of a turncoat. The rhetoric of the industrial hero—"We must feed the world!"—rang out. Things were thought to be getting better in spite of a growing dead zone in the Gulf of Mexico due to industrial chemicals applied throughout the watershed, despite the demise of small towns and rural communities, despite more and more (and ever larger) feedlots.

But things do change.

By the early twenty-first century, talking about ecological agriculture had become easier as evidence of the destructive consequences of industrial agriculture piled up. The reception for talk about perennial polycultures warmed a bit. And with a group of highly qualified young scientists, featuring plant breeders and ecologists, The Land Institute's work expanded. In 2019, The Land had forty-one research colleagues at sixteen universities in the United States and around the world (South Africa, Turkey, Italy, China, Germany, Mali, India, Ethiopia, Canada, Sweden, Uganda, Argentina, Australia, France, Uruguay, and Denmark).

And so here we are now, working cooperatively with some of the land grants. Our first perennial grain, Kernza®, a relative of wheat with the common name intermediate wheatgrass (originally identified as a good breeding candidate by Peggy Wagner, who was on the staff of the Rodale Institute at that time), is out there being grown by farmers and made into beer and other grain products. Still low in gluten, it must be mixed with wheat to get bread to rise, but it is suitable for pancakes, cookies, and other grain products. Our perennial sorghum is grown in Africa. We helped preserve wild perennial rice from the International Rice Research Institute in the Philippines, and that rice is now being grown in China in thousands of acres in paddies, with work going forward for perennial rice that can be grown on the upland slopes. The Chinese, following Vietnam, Laos, and Cambodia, have now started the effort to save their upland soils.

We still have a challenge in making clear that if we stop with perennial monocultures, we will have missed half the point. We want polycultures, plants grown in mixtures. Mimicking the ecosystem, such as the never-plowed native prairie, remains our goal. This is more plausible now that we have the perennials in the pipeline, so to speak, which means we have—pardon me—new "hardware" with the perennials. Now that knowledge coming out of the broad disciplines of ecology and evolutionary biology, which has been accumulating on the shelf for a century and a half, is becoming available. We can begin to more aggressively blend the two cultures—the ecologists who have had the luxury of being descriptive and the agriculturists who have had the burden of being prescriptive—given that our food supply is on the line for the farmer and the agricultural researcher.

It is wonderful to be able to imagine that the primary way of doing business on our agricultural land for the past ten thousand years can change.

One Thing Leads to Another

Some stories have to wander to find the point.
In some stories, the wandering is necessary to get to
the point. In other stories, the wandering is the point.
Wandering around in the past is a good way to figure
out the present. Remember what Faulkner wrote:
"The past is never dead. It's not even past."[1]

Over the Fence Is Out:
Softball Rules at West Indianola, District 93

West Indianola, Soldier Township, Shawnee County, Kansas, some three miles from the Topeka city limits in the Kansas River Valley. From the front doors on the north side of the school I could see the fields of my aunts and uncles and also the house and outbuildings where my grandparents had lived most of their married lives and raised their four children. I could see our own farm.

That two-room country school had eight grades: one through four in the Little Room, five through eight in the Big Room. No kindergarten. The school year lasted eight months, not nine. When spring came, many schoolkids were expected to be on hand to help with getting the crops planted and harvested, or whatever else was needed at that busy time. The distribution of around forty students in those two rooms was relatively equal. We had what we then called "cripples" in our schools due to polio. Jonas Salk had not yet developed his preventive vaccine. Two were children. The third was Mrs. Wade, who taught the grades in the Big Room and also served as principal. The Big Room had a stage that was elevated about ten inches. For the Christmas play, the large doors dividing the two rooms went up.

Two girl students, both younger than I, had difficulty walking because of their polio. One wore braces on both legs, and the other had one leg that was shorter than the other, causing her to walk with her hand splayed between the hip and knee. We took both of these fellow students as they were and made adjustments for their condition. When it came to recess, for example, we had different rules that I am pretty sure we students made up to accommodate their limits, especially when we played softball. From memory, for the one with braces on both legs, swinging the bat made it hard to keep her balance. She was entitled to five strikes. When she hit the ball, someone else ran the bases for her. The older one, with one limited leg, seemed steady on her feet and could swing all right but could not really lean into the pitch, so she got four strikes, and again someone else had to run the bases for her.

We played what was, and maybe still is, called "work-up." With work-up there are no teams. At the beginning of recess, we lit out for home base on the ball field. The first to touch the base batted third, after the two polio-stricken girls. The second one to touch the base batted fourth. After that, touching the base established the position you took: catcher, pitcher, first baseman, then second, third, short-stop, left field, center field, and finally right field. If there were more, then they could fill in out in the field. When someone was put out, everyone moved up in that same order. Sometimes we played with the rule that when someone caught a fly ball, they exchanged with the batter. At other times it was just an out. I can't remember how we decided how it was to be.

We had other rules that weren't regulation. For example, a checked swing (what we called a half strike) counted as a strike. More important, over the fence was out. Jere Bleier could hit a slow underhand pitch and send it over the fence regularly, and this rule restricted Jere to lower-power hitting, leveling the playing field a bit. When Jere did sail one over the fence, he was out and had to go to right field. I have no idea how the rules were drawn up. They were just there, like the school itself. We were mostly left alone on the field. Sometimes the teacher would call the balls and strikes, but not usually. The teachers usually talked with one another. The Little Room teacher was sure to hang around the little kids at the swings and teeter-totter.

Some of the little kids wanted to play ball. When they pitched, they were allowed to shorten the distance between the pitcher's mound and home plate. Now and then a kid would get mad. He or she would insist that he or she had made it to base and was safe. The hue and cry represented a vote, and so in a pout that kid might head to the swing set or the teeter-totter.

We had other games: Pum Pum Pullaway and, with snow in the winter, Fox and Geese.

By the time I made the eighth grade, it seemed that everything had changed. We had a new school, complete with gymnasium. No more softball. We roller skated in the gym and played volleyball and basketball. The charming old two-room school that both my mother and I had attended was gone. My maternal grandfather, as a school board member, had encouraged the community to pay the cost of building that school in just one year—and they did it. But now the old school with its bell tower that called us in from recess was gone. And they added a hot lunch program. No more lunches in paper sacks to be blown up and popped after lunch outside on Friday. No more metal dinner buckets and lunch pails.

I think I was the last to ride my horse to school. But I felt sorry for Poncho standing out there all day, tied up, so I gave that up. My mother, in her day, rode Topsy to the same school and released him. She would remove the bridle, and Topsy would just run back home. In her day (she was born in 1894) there was no concrete highway, no speeding cars. How softball rules were made in her time, she never said.

More needs to be said about the new school that included the gymnasium and newly established lunch program. I started eighth grade in the fall of '49, four years after Japan surrendered in August. The instruments and products of progress were on people's minds, but not so much on the minds of my parents and other older people who had weathered the Great Depression and the Dust Bowl and remained thrifty and frugal.

Thrift and frugality were long in dying for such people, but even I could see extraordinary change underway. When the war started, our family farmed exclusively with horses. My two brothers went to the Pacific to fight the Japanese. While they were gone my father bought a Ferguson tractor with steel wheels, which was all one could get, given

All eight of us in the summer of 1942. *Left to right*: Dwight, Elizabeth, Nettie (Mom), Elmer, Howard (Dad), Margaret, Harley, and Wes (author) in front. Elmer was home on furlough, with Dwight planning to enlist in January; both were stationed in the Philippines. Dad was doing defense work as a carpenter while Harley and Mom did the farming.

that all available rubber tires were needed for the war effort. We still had a team of mules and a team of horses. Somewhere in the midst of it all we no longer had one of the mules and one of the horses, just Bill the mule and Bird the horse. They became the team that we used to pull the wagon as we picked corn by hand. The wagon had a

Author as an infant, riding Bird the horse.

fifty-bushel capacity, and a good hand could pick a wagonload in the morning, unload it with a scoop shovel, have lunch, and pick another fifty bushels in the afternoon. That was a good day's work.

After the war, Uncle Roy borrowed Bill and Bird to pull the wagon as he handpicked his corn. On the stretch through the field the team required no more command than "Get up" and "Whoa." Bill and Bird were kept at Uncle Roy's place all during his corn harvest. While picking during the day, Uncle Roy noticed that going one way, he had a hard time getting them to go, and on the return, which would have been toward our farm, he had a hard time holding them back. They were homesick. This was in the fall, during pheasant season, probably early November.

One day, about 1:00 or so in the morning, old Bird got out from Uncle Roy's corral and headed to her home, our place down the highway. Just as she crossed the road in front of our house, she got hit by a car carrying pheasant hunters returning from hunting in western Kansas. A following grain truck ran over her head. I saw her the next morning, dead along the highway. Before long the dead wagon arrived. The driver tilted the bed, and with a power winch attached to her hind legs he pulled her in and took her to the "rendering works" to be turned into dog food. Now we were down to Bill the mule. Sometime later, someone came through from the South buying up the area's horses and mules and took them back to where I can't say, but I imagined it then to be Mississippi. After the war we got a one-row Woods Brothers corn picker, later a two-row.

One other memory that has to do with dramatic change. During the war, in 1944, Goodyear Tire and Rubber Company began to build a tire factory three-quarters of a mile east of home along the highway. They started with one building on the north side of the road and then added another on the south side. Finally they merged the two buildings, blocking two national highways, 24 and 40, which ran together through this area. This forced a new road to be built at a right angle to go south and connect with another road. The demand for automobiles and trucks, along with their need for tires, was powerful enough to reroute two national highways. On that stretch of old road into town, during the war I could read a sign on the way to Sunday school: "To save tires drive under 35." That part of the highway was now blocked. None of this change seemed right to me. I have been asked several times when I became an environmentalist. My best answer may be that it was then.

Back to the school. Many decades later, thinking about empathy and a sense of fairness, I realized that a kind of grassroots democracy was at work on the West Indianola softball field. No Little League coach was necessary to teach us how to field a ball or to bat. No discipline that I can remember was imposed on the playground by any teacher.

So what was the gain from all the "progress," from the school bus carrying the kids to and from school and getting exercise in the gym? We had a tractor, but Bird was dead, and Bill had gone to the South, where I imagined him pulling a single-row cultivator like one we had, but in some cotton field. I had never seen a cotton field except in pictures in a geography book. Even so, I missed him, and I wondered if he missed me.

I'm not a historian, and I doubt if I will ever be asked when the United States began its decline. I can tell when it began for me. It began in my single-digit years, when Bill and Bird left the farm to be replaced by a small Ferguson tractor with steel wheels, when a beautiful, totally adequate school was torn down and a new one built that included a gymnasium for exercise, and before that when a war effort allowed Goodyear Tire and Rubber Company to block a road that had two federal highway numbers. The decline began when an aggregation of kids with an interest in fairness, perhaps fueled by a postwar growth economy, lost important aspects of everyday rural community life that accommodated young girls who had been stricken by polio. The growth economy really began to take off in those years. What about the elimination of polio, Jonas Salk and his lab that did the research that led to the vaccine, and later the oral form of it? How about that? Of course we are thankful for Salk's discovery. But it's not mere nostalgia to point out not only what was gained but what was lost with all that progress. Morris Berman, in *Dark Ages America*, reminds us that culture is a "package deal."[2]

And how about what could have been a legitimate home run that went over the fence? How were the rest of us expected to field a ball that went out of bounds? As we think of social justice in the period of reduced fossil-fuel burning facing us, what is wrong with the idea that over the fence is out? I want to think about that.

Sharon Stays at Home, Mostly

In late August 1954, I drove my robin's-egg-blue 1946 Chevy coupe to start college at Kansas Wesleyan University in Salina. I arrived a week or two ahead of classes for football practice and checked into my room in Schuyler Hall, an old building with hallway floors that here and there had a wave-like reality. Some of the doors to the rooms had to be trimmed at an angle in order to shut. It was a small school, with an enrollment of less than three hundred, that became accredited sometime before I graduated.

I was eighteen years and two months old, and when I arrived I went by Sharon, my first name. There was a Topeka-area singer named Sharon French, a man my mother loved to hear. She also liked the name. It seems that there were not many people with that name before that period, and most who had it were male. Now, of course, most Sharons are women. Growing up in that rural community, the name was not seen as unusual and was simply who I was. Only now and then did someone who did not know me feel it necessary to question or comment. However, during the first week of football practice it became apparent that there would be more questions. My middle name was the less confusing Wesley, and so I told Coach Gene Bissell to just call me Wes. He readily complied, and that is who I became.

What had served me well for over eighteen years was not totally discarded. I am still Sharon to relatives and friends back home and Uncle Sharon to fifteen nieces and nephews. My driver's license says Sharon Wesley, as do my passport, tax returns, bank statements, etc. Sometimes I switch to S. Wesley. When official forms or special deliveries come with the name Sharon, there can be some momentary confusion.

Wesley is at once a family name and a Methodist name. My grandfather was Charles Wesley Stover. Charles Wesley was a hymn writer and the brother of John Wesley, the founder of Methodism, a name given by others who noted that the movement had a method by which

Author's maternal grandparents and their children. *Left to right, top row*: my uncle Harry, my mother Nettie, and my uncle Roy. *Bottom row*: Grandmother Ann Pearson Stover, Aunt Ruth Stover Dister, and Grandfather Charles Wesley Stover.

someone who is not yet among the "elect" can become elected for salvation. Here is part of that story.

The story of Wesley's conversion sometimes begins in 1735, when John and Charles Wesley were on a ship bound for the Georgia Colony. Some of the passengers were English, and some were Moravian missionaries from Germany. There was a ferocious storm. The mainsail shredded, the decks flooded. While many English passengers screamed, worried that the sea would soon claim them, the Moravians were calm and sang through the storm. They were not afraid of death. John Wesley wrote about that earlier experience in his journal. Then on May 24, 1738, in London, he felt that he had been saved:

In the evening I went very unwillingly to a society in Aldersgate Street, where one was reading Luther's preface to the Epistle to the

Romans. About a quarter before nine, while he was describing the change which God works in the heart through faith in Christ, I felt my heart strangely warmed. I felt I did trust in Christ, Christ alone, for salvation; and an assurance was given me that He had taken away my sins, even mine, and saved me from the law of sin and death.[3]

Wesley had been struggling with his lack of faith but kept at it through study. He had a Method. Such a way of doing business with the Lord became well known, and Wesley's Method caught on and was brought to America.

A major disperser of Methodism in America was Bishop Francis Asbury, who made circuit riding a common way to spread the Gospel. The riders would travel by horse from town to town. Asbury worked the eastern United States pretty hard—particularly, as I understand it, in the mid-to-lower Appalachians—and he must have featured some of Charles Wesley's songs. My grandfather was born in 1855 in the Shenandoah Valley of Virginia in a community that was mostly Lutheran, though his parents and grandparents are buried in a Methodist cemetery there. I imagine some circuit rider must have come through, featuring Charles Wesley hymns, and that was how he got the name six years before the Civil War began.

So there you have it. Sharon is a biblical name, after the Plain of Sharon in Israel, along with the "Rose of Sharon" mentioned in Song of Songs. Wesley is a denominational name, and Jackson is secular—three bases covered.

At the risk of being ungrateful to my namesake, most of the music for those Charles Wesley hymns is horrible. Some of the words don't quite do the trick, either, but there are some with useful insights. Which brings on another digression.

In 2018, I went with two of my football buddies from college days at Kansas Wesleyan to a memorial service for Butch, one of our teammates. The service was held in a beautiful Methodist church where

Butch's grandfather had been pastor when Butch was growing up with a single mom. The family had scattered. Maybe only a dozen people besides the three of us were at the service to pay last respects in a formal way. Butch had been a veteran, and so there was a military funeral. There was a lot of waiting before and even during the service. What are you to do when the only book in the back part of the pew facing you is the Methodist Hymnal, along with envelopes for the offering—not even a Bible? Well, you thumb through the hymnal and read some verses of the old hymns. I happened upon a few that I thought might be of use someday.

Attendance in that small rural church was going down, along with the population of the town. I thought, "I could use something out of that hymnal now and then that no one else will ever think to consider, and there are plenty of them around." Were the pastor around, I could have asked to buy one. Most of these books would likely go to a dumpster when the church finally closed. If I took this hymnal, that would be stealing, but who would care? And then I thought something else: "What if there was a Bible? What if I stole a Bible and got caught? Wonder what anyone would say?" Taking a Bible is spreading the word, isn't it?

There was no Bible and so no temptation there. But that hymnal? Well, I didn't take it, but I told my friends driving back that I wished I had one of those Methodist Hymnals. My friend Bob was also raised as a Methodist and is married to Karen, a Methodist who is also a friend of mine. He told Karen of my wish, and she found a hymnal and gave it to me. I think she got it from her church. All their hymnals had been removed from the pews and were stacked in a church closet.

Kansas Wesleyan is obviously a Methodist college, but I went there primarily to play football and run track. Nevertheless, with my emphasis on Wesley and having been baptized in the Methodist Church at the age of twelve, the age of Jesus when he stayed behind in the temple, I had credentials few enjoyed. It served me well even after the mid-1960s, when I lost "the way."

Here is how that happened. I had been leading an adult Sunday school class after I returned from my PhD program in North Carolina to teach at Kansas Wesleyan. On this particular Sunday, I asked the small gathering of maybe eighteen people, mostly young couples, with a child or two, to go through the Apostles' Creed with me, give a show of hands when they agreed with the creed's claims, and "interrupt when you feel like it."

Like most people in my age group back then, I had learned the creed by heart, the same way I had learned the books of the Bible and the order of the presidents of the United States. It was no big deal to be able to recite all that. So I began, "I believe in God the Father Almighty, maker of heaven and earth." Hands went up—all of them. But, cruising along, I eventually came to "born of the Virgin Mary." Well, that was a showstopper right there. Couples were looking at one another and around the room, uncertain whether to signal their skepticism. I continued on to "suffered under Pontius Pilate, was crucified, died, and was buried." All hands up on that as the historically accurate part, and then I came to "the third day He arose from the dead." Well, now some hands went up, some stayed down, and some were partway up, with more looking at one another.

I saw no need to finish. I simply said something like "Dear friends, this is what our membership is supposed to believe. Are we untruthful to ourselves or one another?" After nearly all had left, there was one woman remaining who was in tears. I asked what the matter was. She had recently been advised to get into a church group, and now I had taken away everything. I was sorry for her. I told the pastor, and he immediately said, "You should not be teaching that class. In fact, you should not be in the church." Doubt is supposed to be a part of faith, but he apparently calculated that my doubt-faith balance had tipped in the wrong direction.

I counted myself as fired that day from the Methodist Church. I had been a member for nineteen years, over 60 percent of my life at that time. That ended that run. These days, I count myself a five-eighths Christian—a Sermon on the Mount and "he who is without sin cast the first stone" kind of Christian.

What do I believe beyond those inspired words of Jesus? What do I hope for? Well, I do hope—and perhaps hope is a form of prayer—that our knowledge of the journey from our stardust origins to planet formation, the journey from minerals to cells, and the Darwinian selection story, which includes the new epigenetics, will inspire us all. This is verifiable information that the earth is—as the old hymn "O Worship the King" put it—our maker, our defender, and, with proper ecological restoration, our redeemer. It's possible that all this knowledge can deepen our reverence for our Earth and help us see new opportunities to be participants in the creation, in the living earth.

Most of us try to avoid death as long as possible, but we all know nothing holds still—it's a dynamic universe, not a static one. Birth and death are part of the big equation, down the line for whatever exists at any moment, even for the rocks at our feet and the stars in the sky.

I continue to read the Hebrew scriptures and the New Testament and take both seriously, but in my own way. By the pastor's standards, I am pretty certain that won't get me back into the Methodist fold as he understood it. I can't say for sure whether John and Charles Wesley would approve if they were alive today. Had they known what science now tells us, maybe they, too, would have moved on to accept the idea that we have all been cycled through a supernova at least twice—though, of course, as tiny pieces.

Why Can't I Pick Up Dug Potatoes Fast?

I learned a lot in Uncle Harry's potato fields, where this story begins. In very late June and early July, boys and a few young men would handpick the harvest, some years as many as two hundred acres. One full wire basket was half a bushel. Two full wire baskets would get dumped into a burlap gunny sack, and on Friday afternoon all diggers were paid in cash—ten cents for every bushel. My cousin Danny was a year and a week younger than I, and he almost always picked a few more bushels than I could manage. I mentioned this to Uncle Harry once, and here are his words, meant for comfort, I am sure: "Art always had the fastest hands in the cutting cellar."

First, who was Art? My mother had three siblings: Harry, the farmer, who had a diverse operation; Roy, who also farmed, was less expansive, and helped Harry at times; and Ruth, married to Art. So Art was my uncle by marriage to my mother's sister. They were Danny's parents and lived on the adjacent farm east of us, my grandparents' homestead.

What's this about the "fastest hands in the cutting cellar"? For those who never thought about it, the part of the potato we eat is an underground stem, not a root. The potato "eyes" are belowground buds that will sprout and grow as they draw upon the potato's energy resources to poke through the ground and produce the green stem and leaves we see above, there to capture the sun's energy and send much of it to be stored in those underground stems called tubers. What we might call the perennating tissues (what stores energy to make it through a winter) for potatoes are all those inflated underground stems.

Seed potatoes are actually just sections cut from ordinary potatoes. That work was done in the cellar each winter for planting in March, usually sometime around St. Patrick's Day. Uncle Harry would refer to the "sign of the moon" to determine when to plant. I still wonder if he took that seriously or was agnostic on the subject.

Winter in the cutting cellar involved up to a half-dozen men, mostly neighboring farmers, who would sit on various kinds of stools, such as five-gallon buckets turned upside down. They would hold a potato in one hand and with a sharp knife in the other cut and turn the potato to be sure they captured at least one "eye." The cutting with the eye had to be large enough to fit within a man's partially clenched fist, which would make the cutting big enough to sponsor the necessary growth to push through the soil, become green, and capture sunlight. "Eyeing the potatoes" was winter-in-the-cellar work.

There was clearly an art to all of this. Finding the eye, turning the potato to inspect the size, and with the same motion a deft twist of the wrist to turn one potato into two or three seed potatoes. Some men were faster than others, and according to Uncle Harry, his brother-in-law Art was the fastest. If Danny had inherited his dad's fast hands, then my slower hands were the result of genetic limitations, not a failure of character. What a relief!

The potato cellar did not freeze, but it was plenty chilly. The men dressed for it, though, and were sufficiently comfortable to tell stories and jokes and talk about last year's crops, what the almanac said, and the state of political affairs. They were all Christians of various degrees of faithfulness, mostly Protestants with one Catholic.

Potatoes were planted in mid-March, and some three and a half months later it was time to dig them. That's where I came in, to help with the harvest. I started on my knees, picking potatoes off the ground along with fifty to a hundred other pickers.

Potatoes were planted in rows. At digging time, a two-row potato digger went down the line, lifting the soil and potatoes, with the cutters going deep enough to capture the potatoes from below the ground. The dirt and potatoes would land on an endless moving chain that bounced and shook the dirt off before discharging the now exposed potatoes onto the ground. Those boys and men picked them and put them in a wire basket and emptied the basket, when full, into a burlap gunny sack. Done twice, that's a bushel, weighing sixty or seventy pounds.

Each person was assigned a section defined by the distance between stakes. That distance was initially determined each morning by two of Uncle Harry's sons, my adult cousins, Charlie and Joe Stover. As the sequence of pickers was being established by the setting of stakes, an older boy, age seventeen to twenty, would follow with punch cards. He would write the picker's name on two cards, one for the picker. The other cards were on a ring that he carried. At loading time, each bushel was recorded on both cards with a punch. The number of punches determined how much the picker would be paid.

Now comes a potential source of trouble. With each round of the digger, the stakes would be moved by the pickers. Even though Charlie and Joe made the rounds, they couldn't be everywhere—and they didn't need to be except when they saw that a section was not being well picked. Then they would shorten the section. The gain would go to one side or the other. Charlie and Joe wanted to be sure the pickers did a clean job of picking. Most boys wanted to have a longer section, which meant more potential to make money. While Charlie and Joe were otherwise occupied, which was usually the case, the stakes were moved by the pickers. So, after the digger passed, there was a close eye on the one moving the stakes. There could be arguing, but I never saw a fight.

Now a word about the trucks and the bag loaders. As the truck would come by, two boys, usually fifteen to seventeen years old, on the ground would heave the potatoes four feet up onto the flatbed to be received by two more boys of similar age on the truck bed. It was just before the loading that the counting of sacks and number punching happened. Cheating was minimized. One issue was dirt clods, which weigh more than potatoes. The loaders would often catch this, and sure enough, upon examination there would be clods in the sack. Or if a burlap sack had its open end hanging over too much, the punch-card man would have the contents of the sack poured into a wire basket and then dumped into another sack. The remaining contents of the initial sack would then be emptied into the wire basket, and if the sack was not full or nearly so, it wasn't counted until it was full the next round. It took very little effort on the part of the one carrying the punch, or one of the loaders catching the extra weight, to ensure what appeared to be a high level of honesty among the pickers. With two seasons of watching and participating in all of this, I got a sense of the pickers' competence and the human capacity for fairness.

At any one time, there could be kids from four or five communities, more or less randomly dispersed in those fields. No segregation in the fields, though the living arrangements in Topeka were segregated. There was a Mexican community, distinct from the African American community. There was a place called Little Russia, which produced "RUE-shun" kids, in our vernacular, along with a few Latvians and Poles. The African American kids played "the dozens," which was a real education for white farm boys like me.

A few of the fields bordered the Kansas, or what we usually called the Kaw River. At lunchtime many of us pickers would plunge into the river to cool off with a swim. No segregation there, either. While Topeka communities were not integrated, the Kaw River by the fields was. It never occurred to me until adulthood that the various tribes likely had been swimming there for thousands of years. But they were not there now, and they had been erased from our minds. Kansas had achieved statehood ninety-one years before. Fool Chief's Village, on the upper terrace a mile or two from where we dug potatoes and two miles from my home, had been covered by a major flood in 1844. The Kaw Tribe relocated a few years later to a site near Council Grove, Kansas.

(A note on names: The Anglo name "Kansas" comes from the Kaw Nation, sometimes called the Kanza or Kansa, a federally recognized tribe.)

After two seasons of picking, I was given the job of water boy. Two wooden barrels, one holding fifty gallons, the other two-thirds of that, each with a spigot, were mounted on a horse-drawn artillery cart. Carts like this were used overseas in World War I, and several were returned to Fort Riley after the war. The army didn't need them anymore, and not because that was supposed to be "the war to end all wars." Instead, the acceleration of the still somewhat early stages of industrial farming had increased the fossil-fuel-powered industrialization of nearly everything. Those artillery carts were made available, I imagine at auction, for whatever purpose anyone, maybe especially the farmers, could use them for. By the way, Uncle Harry's old artillery cart still exists in the family. One of his grandsons has it, along with other farm machinery treasures housed in one of his sheds from that midcentury and earlier period.

My job was to see that all the workers had water. The water wagon was pulled by Old Bob, a large draft horse who was about eighteen years old then, usually gentle as they come. Bob was a gelding, which means he had been castrated, and therefore more or less predictable in his ways. There were limits to Bob's patience, though. Some of the boys in the field would torment him by throwing clods and potatoes.

I got paid seven dollars a day as a water boy, and since I never could pick seventy bushels of potatoes, it was a step up for me. But when the potatoes and clods started flying, that's when I earned my pay. Most were thrown by boys who were on their knees. I needed to keep a tight hold on the reins and wait until the barrage was over before moving on. My hands may not have been as fast at picking potatoes as Danny's, but my hands and arms were stronger, which was important to keep Bob from bolting. I knew there had always been genetic differences among creatures. This was a help because whether hoeing strawberries or putting up hay, it was a matter of character if one fell behind or could not "pull his own weight." It was a sign of even better character if you could do more. But it's good to remember that we all have different capabilities that are, at least in part, not about our character.

Uncle Harry's potato fields were seldom more than a mile or two from home. I rode my bike there and back. At home there was the milking, feeding the hogs, gathering the eggs, then the bathtub to wash off a lot of dirt, and finally supper. I didn't have to do dishes when I worked away, so it was straight to bed, a deep sleep, morning chores, breakfast, and off again on my bike.

In 1950, I was fourteen and really wanted to use my body. I asked to be a loader. Loaders had to be strong enough to throw sixty-pound sacks onto a flatbed truck or stack them on the bed high enough to maximize the load but not so high that they might fall off on the trip to Uncle Harry's potato shed near town by a railroad spur. There the potatoes were sorted, sometimes washed, and put into sacks to weigh a hundred pounds, which were then sewn shut using a large needle. Then they were loaded onto railroad cars to be shipped to places like Chicago, with only ice to keep the load cool.

One-and-a-half-ton Ford trucks were the standard in the potato fields. Uncle Harry had three: a red-and-white '35, a black '36, and a green '47. The two older ones were flathead eights. The '47 had six cylinders. At digging time, he borrowed Ernie Whiteman's '41 flathead six. Harry's brother, my uncle Roy, had a '48 flathead six that, even when fully loaded and on soft ground, wouldn't overheat. The two older ones, especially, had their hoods and grills off and carried lighter loads. The beds of all five were about four feet off the ground. Why do I have such a clear memory of those trucks? On the farm, we tended to pay close attention to our tools.

These trucks did more than haul potatoes. In the early morning, three of them would make the rounds through North Topeka and pick up kids at designated stops to be trucked to the digging site. Some would sit on the edge of the bed, legs dangling, or even on the front fenders. OSHA did not exist then.

I grew up a little bit in the potato fields, and nostalgia fuels my regret that those fields are gone. I know of no place in the Kansas River Valley now with potato fields of such a scale. Uncle Harry's was once one of many such operations. His farming operation supported four households from time to time, but industrial farming methods reduced the number of people that his farm operation could support. The speed of that change needs to be noted here. Within a twelve-year period, my parents' farm went from horses to steel-wheeled tractor to pneumatic tires to no horses. The 1951 flood, the largest since 1844, must have set back nearly every farm family in the valley. That said, I sometimes wonder if the flood actually accelerated the change. The possibility of flooding had in the past set limits. More on this later.

After the '51 flood, my potato field days were behind me. The summer of '52, I worked on a relative's ranch in South Dakota. The summer of '53, before my senior year of high school, I got a job as a welder at Topeka Foundry and Iron Works, where, among other jobs, I welded the beams to be installed for a new gymnasium in a nearby town. The summer of '54, I went to work at the Henry Manufacturing plant in Topeka. I became a journeyman welder that summer and joined the union, which required a swearing-in by the shop steward, who, while holding the Bible, switched his cigarette from his mouth to his hand and read whatever it was I said "yes" to. This was not Methodist, not even Christian, I thought, but we got it done.

There was good money to be made at the foundry and even more at Henry's. From the foundry I made enough to buy a car. At Henry's I made enough for my first year of college, so I needed to work only two hours a day in exchange for two daily meals at a local restaurant.

I mentioned earlier that change was fast coming. You could see it in the technological array. At Henry's I built the arms for front-end loaders for tractors and the buckets to be attached to the arms. I built parts for backhoes. The backhoes were more often called "grave-diggers" early on, but that designation eventually faded.

While welding parts for the gravediggers, I would think of Zeke
Evans. Zeke was the sexton at Rochester Cemetery near Topeka, the
oldest graveyard in the county. Lots of pre–Civil War history in that
ground. My grandparents and many other relatives are buried there.
Welding those parts for gravediggers made me think of Zeke, aware
that those who followed would no longer be so dependent on his pick,
shovel, and sharpshooter (a narrower shovel used to go deep) to bury
anyone. I thought I was doing something good and proper to help
Zeke and his like.

Those same front-end loaders would also dig trenches for piping
and replace human ditchdiggers, but I wasn't thinking about them. I
knew Zeke—men in my family would often help him cover a grave at
the end of the casket lowering—but I had never known a ditchdigger.
All I knew was that being a ditchdigger was considered just about the
lowest form of labor. Kids were admonished to do well in school "so
you don't end up a ditchdigger." Such labor was done with shovels
then. It was difficult and poorly paid and, as such, low in status.

When I first started welding, I paid little attention to the fumes
coming off the welding rod. Welding was outside at the foundry and
inside at Henry's. After the first year there I saw that those fumes were
going to my lungs, for at the end of August, all through early football
practice, I had hard evidence. When we ran wind sprints, I would
cough up black mucus. By the time we got through conditioning and
to the first game of the season, the mucus was no longer black. The
second summer at Henry's, I was on the night shift, and during lunch
break, sitting outside on the grass with the other welders and machin-
ists, I thought of the "old-dog welders," as they called themselves, and
wondered what their lungs were like and how short their lives would
be. I have wondered if I am still alive because the wind sprints I ran to
get ready for football cleaned out my lungs.

What about all of that dust associated with farming? My dad,
brothers, and farming cousins "ate a lot of dirt" and lived long. The dirt

from those fields was not industrial dirt coming off a burning welding stick. Dust pneumonia did lead to death for many during the Dust Bowl years out west on the Great Plains, but that was an exceptional period. There was no way to escape the dust, even in the houses.

Through this fast-change period, I managed to sort of keep up with what was happening on the various farms. You could see some of it in the fields from the road. My nephew Charlie Wooster tells me it was in 1954 that Uncle Harry bought another International Farmall M tractor, this time with TA added. The "TA" stood for torque amplifier. The advantage of the TA was that when any piece of equipment met enough extra resistance, such as coming upon hard ground, the tractor would automatically shift to a lower gear.

In 1961, Uncle Harry bought a potato digger that would spill the potatoes onto his '35 and '36 trucks. Using it required the truck beds to be removed, and a container specific for the job was added to the frame to receive the potatoes being discharged from the side of the digger. This technology was the final blow; no more kids digging potatoes on their knees. Besides that, other parts of the country were growing more potatoes. The wonder of all these new features was likely talked about in the potato cellar. But such talk is no longer available. The technology first changed the talk in the potato cellar and then silenced it altogether.

I earlier mentioned the '51 flood as a possible accelerator of change. After that flood, Congress and the US Army Corps of Engineers got busy. Protests broke out when it was mentioned that towns and graveyards would have to be moved. "Big Dam Foolishness" appeared on signs. Many older people like my dad, having grown up in an area that flooded frequently, noted that the gift of fertility in the valley came at the cost of an occasional flood that brought those nutrients. "The valley belongs to the river," my dad would say, an idea he shared with others. The protesters lost. Towns and graveyards were moved to higher ground. Cabins were built around the lakes. Now people could fish, swim, water ski, boat, and invite friends out for barbecues while the ecological capital on the lake floor lay mute and useless.

The dikes on the Kaw are higher now. And with several major dams added upstream, will we ever have another flood like the one in '51? Sure, but when? The dams are accumulating soil due to the erosion from the agricultural fields. When will they become useless? In the meantime, with the perception of protection from flooding, both housing and commercial development near Topeka have increasingly covered the rich soils of the Kansas River Valley with concrete. On one of the biggest former potato fields, parents now bring their kids to compete against one another and learn the skills needed for soccer—all of this covering soil that would be useful for future food. On other former agricultural fields, boys and girls swing bats, throw balls, and run bases—all developing their Upper Paleolithic skills that were, quite literally, lifesavers back in our preagricultural days.

So what are we to make of this slice of history? In the potato fields, kids picked up potatoes brought to the surface with fossil-fuel-powered technology in the form of a tractor and digger. Potatoes lying on the surface for them to pick came out of agriculture. Yes, it was a kind of gathering, but was that job any sort of analog of the Upper Paleolithic gathering-hunting era? They were not digging the potatoes out of the ground with a stick. The potatoes appeared on the surface for those Topeka town kids by the magic of fossil fuel and technology. But they were on their knees most, though not all, of the working day, and it was usually hot. My sense was that they would not be there because it was fun but because they thought they needed the money, or perhaps because their family said that they needed to "pull more of their own weight." What we know is that the technology was rapidly changing until the time when potatoes were dropped onto a container using fossil power. I haven't checked, but I wouldn't be surprised to find that the only potatoes that are grown in the Kaw Valley now are in people's gardens. Technology and specialization win again.

This reflection on the potato fields suggests that I respect those who work hard and take some pride in the work. Hard to argue with that. But could it be that the idea of hard work as a virtue is cultural, the consequence of agriculture and the Fall? Anthropologists tell us that gatherers and hunters put in a shorter workday to feed and provision themselves than we do.

Could it be that, once people were in the fallen world, a strong moral code was necessary to keep them working hard, perhaps because such effort isn't part of our evolutionary history? How and when does a sense of oughtness begin to wither? In an affluent society, more kids spend more time kicking, running, throwing, and hitting a moving target with a stick than picking up potatoes or hoeing. Are they now back home in the Upper Paleolithic, more like the original Garden of Eden, where there was harvest without thistles, thorns, and sweat of the brow? That may be true if we assume that Eden was no garden in the first place but a gathering-hunting paradise.

A Private's Supper on the Oregon Trail near Topeka

I'll start with a digression, with a promise to get to the posted title of this story somewhere in between other digressions that are sure to follow.

On August 10, 1821, Missouri became a state, and the following month Mexico won its independence from Spain. The United States was full speed ahead on Manifest Destiny, looking for opportunities.

The Missouri River runs on a general trajectory from the northwest, starting near Three Forks, Montana, southward and east to a point where it makes a sharp turn eastward to flow across the state of Missouri into the Mississippi a bit north of St. Louis. At that sharp turn, the Kansas River, or what we often call the Kaw, goes into the Missouri from the west in Kansas. This is an auspicious junction in many ways, made especially auspicious at the time because some 750 miles to the south and west is Santa Fe. Southward from there is Chihuahua and a few other centers of economic interest to an expanding country not yet half a century old.

In 1803, scarcely eighteen years before Missouri statehood and Spain being kicked out, Thomas Jefferson struck a deal with Napoleon that we call the Louisiana Purchase. The United States now had the deed to an additional 828,000 square miles at a cost of $15 million. In today's dollars it would be over twenty-three times that.

It didn't take long for a few, at least, to contemplate economic opportunity. In the early 1820s, a trader named William Becknell had begun to lead his teams from near where the Missouri turns at the junction of the Kansas River. He didn't follow the Kansas River but headed west and southward from Westport. Now part of Kansas City, Westport is considered quaint these days because it is loaded with frontier history. Like many other historic places, it has become a good place for fine food, wine, coffee, and locally brewed beer. That route became the Santa Fe Trail, which operated until 1880, when it was replaced by the railroad. It was mostly a trail for commerce; people looking for a new place to put down roots generally didn't head that way. Trade featured wool fleeces and woven goods from the East. Mules and silver came from Mexico to Missouri, making Missouri famous for its mules. The University of Central Missouri's men's athletic teams are called the Mules.

As a nation working from the assumptions of a government run by white men, the United States wanted to protect "our" interests, which meant it needed a substantial fort on the other side of the Missouri, in Kansas Territory. In charge of building that fort was General Henry Leavenworth, who had fought in the War of 1812 and had been involved with conflicts between the US government and the native populations. The first identifying name for the new fort was Containment Leavenworth, dated May 8, 1827.

When we think in today's dollars about that $300 million that we spent less than half a century after winning our independence from England, we now had control of not only the best lands on the richest continent in the world but also the major rivers of the continent and harbors with deep drafts to the east as well as the south and west, all the way to New Orleans. The slave population was making the nation richer. All that was left, really, was to continue the de facto genocide of the native populations. The US government got help there, too, from the diseases settlers brought with them. This sounds cynical, but it may not be cynical enough. The slave population had been increasing in the South since 1619, for more than two hundred years before the construction of what is now called Fort Leavenworth, so there was the growth of the economic base from slave labor. What would our country be without the near-complete extermination of the native population and well-ordered slavery? It's a question the dominant culture has long done its best to ignore.

Fort Leavenworth is 250 miles west of the Mississippi on the Missouri River. Take another hop 300 miles west and we are at midcontinent. At this point, our territories were already beyond the geographic center of what would become the Lower 48.

So now we have a Santa Fe Trail going west and south to serve commerce and our military. We have Fort Riley west of Leavenworth in place twenty-six years later, where the convergence of two streams from the Great Plains, the Smoky Hill and the Republican, form the Kansas River.

The trail from Fort Leavenworth to Fort Riley was around 130 miles. The trails were somewhat braided, but near Topeka, some sixty miles from Fort Leavenworth, a military road met one branch of the Oregon Trail, still some seventy miles to Fort Riley.

I realize that a complete description of the convergence of those early trails involving commerce, military transport, and travel to Oregon could become tiresome. Just know that there was a farm near Topeka where all three converged. That farm happened to be a convenient place to camp in a pasture, at least in the last two decades of the 1800s, and probably before as well as later. That was my grandparents' farm, and here is where the story begins.

I had heard stories many times about the military camping on the corner pasture of the property. I had also heard tales about settlers coming back from the West in covered wagons to visit family in Missouri or somewhere, the speculation being that they were unable to confess that they had been dried out on the homestead to the west. I'd heard about how soldiers would come to my grandparents' house to "buy their supper." Army rations weren't the best, one can imagine, especially for a moving cavalry.

And now a story about a private and an officer who walked independently some three hundred feet north to buy their supper from my grandparents. The private arrived first to inquire and was accepted. Some minutes later the officer arrived at the house and asked to buy his supper, too. His request was also granted. Now, supper was soon enough ready, and it was time to be seated, with the blessing of the

Farmstead of Stover grandparents, pictured within the yard's fence. To their right are Roy and Harry with two teams of horses. This is where my mother proposed to my father in a buggy. The fence my grandfather is leaning on faces east and was part of the road the military used from Fort Leavenworth to Fort Riley. To the south is the pasture where travelers often camped at the intersection of the military road and a branch of the Oregon Trail.

food to follow. There was only enough space in the dining room for the family and one soldier. Someone would have to sit at a small table in the kitchen. The private stood aside to allow the officer the spot at the dining table. At this point, my grandfather, Charles Wesley Stover, told the private, "You were here first. He can sit there," meaning that the officer would be the one seated off to the side.

I didn't know my maternal grandparents. My grandfather died eleven years before I was born, my grandmother a year after my birth. I wish I knew more about what they were like, especially when I ponder how it was that a private outranked the officer for a position at the family table. My grandfather was born in 1855 in the Shenandoah Valley of Virginia. The Civil War was 1861 to '65, making him six when it started, ten when it ended. His father would have been expected to serve in the Confederate Army, but he hid out during the war and sneaked back home from time to time. Here is where a bit of history comes in handy.

In August 1864, Major General Philip Sheridan came through the Shenandoah on a campaign of "burning the valley out." The goal was to destroy the "Breadbasket of the Confederacy" and break the will of the South. Grant had given specific instructions to "eat out Virginia clean and clear as far as they go, so that crows flying over it for the balance of the season will have to carry their own provender with them."[4] On August 17, Sheridan reported, "I have burned all wheat and hay, and brought off all stock, sheep, cattle, horses, &c, south of Winchester."[5] They went south of Harrisonburg and raided as far as Staunton and Waynesboro.

My ancestors were directly in that path, for their village was Fort Defiance, near Mt. Sidney. My great-grandmother was alone on the farm with the children when the Union soldiers came through, using incendiary bullets designed to start fires. The story goes that as one of the soldiers raised his weapon to set the barn on fire, my great-grandmother touched his arm and said, "Son, that won't be necessary," and managed to save the barn from being burned. That barn was still there in 1917. Maybe the Union soldier didn't have it in his heart to follow Sheridan's orders, and her gentle reminder to him of his humanity was all that was needed.

A little perspective here: All four of my grandparents were born before the Civil War. One grandmother, from near Gettysburg, shook Lincoln's hand. The other was born in the Territory of Kansas at the time of "Bleeding Kansas" (1854–1861) before Kansas became a state. Not far from Lawrence, she and her family hid in a cornfield on August 21, 1863. That's when Confederate raider William Quantrill and his 450 men ravaged and burned Lawrence and killed more than 150 people. Nearly all of the dead were males, some in their teens.

Now back to the supper for the family and two soldiers. My grandmother, the one responsible for that meal, was the one who hid in the cornfield. Before she was born, her father was with John Brown and other abolitionists at Black Jack Creek on June 2, 1856. Outnumbered, they still captured a proslavery contingent in what some have called the first battle between free-state and proslavery forces.

How many suppers, dinners, or breakfasts were sold along the trail, we'll never know. By the turn of the century, the practice would have pretty much ended. America was on the move. Railroads were moving traffic, and little towns were springing up here and there. No need to buy a meal in a farmhouse. The changes also brought an expansion of information, and newspapers became easily available. There was a great increase in knowledge, and people could keep up on what was going on locally and around the country, even in other parts of the world. Many colleges, especially church-related colleges, attracted the children of homesteaders.

So it was not surprising when I learned that Charles Wesley Stover, my mother's father, had taken the horse and buggy from the Kansas River Valley north of Topeka—date unknown to me—all the way across town to the south side of Topeka to Washburn College. There

My great-grandfather Robert Hall Pearson arrived in
Kansas in May 1954, a few days before the ratification
of the Kansas-Nebraska Act. He was with John Brown
at Black Jack Creek near Baldwin. Later he bought the
property where the battle took place. The site is now
called the Robert Hall Pearson Memorial Park.

he heard a lecture about Charles Darwin and the radical ideas in *On
the Origin of Species*, published in 1859, two years before Kansas became a state. Washburn had been founded in 1865 as a Congregational
school called Lincoln College. It was renamed Washburn College after Ichabod Washburn, who made his fortune with Washburn and
Moen Wire Works and contributed $25,000 in 1868.

A historical footnote: Abraham Lincoln and Charles Darwin were
both born on February 12, 1809. For years now, I've had fun pointing
out that coincidence when discussing evolutionary theory in public
talks. "Lincoln and Darwin, born on the same day," I would say. "One

went on to become the Great Emancipator, and one became president of the United States." My point, of course, is that Darwin freed us from many misconceptions, especially the notion of the divine creation of species.

Well, it appears my grandfather did not agree. My mother told me that her father came home from that lecture furious about what he believed to be nonsense. Given enough time, would he have come around to the evolutionary point of view like his geneticist grandson? Hard to say, given that there's evidence he was something of a mystic, believing in the power of some minds to make predictions. That conclusion comes from one of the many stories I heard about—and now here's another name to keep track of in the story—Coleman, who had been enslaved before the Civil War. Coleman occasionally worked for my grandfather, doing seasonal field work and sometimes odd jobs such as mending harness. When Coleman came by, my grandparents would immediately welcome him into the home, except one time when Grandfather met him at the door and announced, "I'm not going to let you in, Coleman, until you tell me where my brother Milton is."

His youngest brother, Theodore Milton, went by his middle name. He had left the family and gone out on his own in his early teens. Milton traveled around the world more than once in his lifetime, if I remember the telling right. He served in the Boer War in South Africa and was something of an expert horseman, favored by Queen Victoria, who asked for him specifically to ride the hurdles for her in England.

"Where is he, Coleman?"

Coleman lowered his head, one hand on his forehead, and with eyes closed was quiet a good long while until finally, with eyes still closed, he pointed westward over his shoulder and said, "He's that way, and he'll be here soon."

"How soon, Coleman?"

"Maybe a month."

It had been something like five years since my grandfather had seen his younger brother. My grandfather seemingly believed that Coleman had some special gift. Did Coleman simply make an educated guess and hope to be right? Whatever the case, Milton did arrive shortly after. And he came from out West.

A bit more about Milton, the youngest of my grandfather's full sibs. There were eight children in the family. The only girl from that family was their sister, Ida Stover, who married David Eisenhower in Lecompton, Kansas. She named her youngest child Milton after her baby brother. After the elder Milton's global travels ended, he settled in Myrtle Point, Oregon. When his nephew, Dwight, ran for president, Uncle Milton told an interviewer that he didn't think he would vote for him. His reasons are lost to history.

My Rural Life

Living with less cultivates virtues, though a bit of vice
is inevitable wherever one lives.

The Matfield Green Women's Club

In the early 1990s, I lived on and off in the small Kansas town of Matfield Green in Chase County, population fifty-six. The Land Institute had a presence there, buying the abandoned elementary school and a few other buildings in the spirit of rural revival, and I acquired a few abandoned houses myself. At work on them, I had great fun tearing off the porches and cleaning up the yards. But it was sad, as well, going through the abandoned belongings of families who had lived out their lives in this beautiful, well-watered, fertile setting.

In an upstairs bedroom, I came across a dusty but beautiful blue padded box labeled "Old Programs—New Century Club." Most of the programs from 1923 to 1964 were there. Each listed the officers, the club flower (sweet pea), the club colors (pink and white), and the club motto ("Just Be Glad"). The programs for each year were gathered under one cover and nearly always dedicated to some local woman who was special in some way.

Each month the women were to comment on such subjects as canning, jokes, memory gems, a magazine article, guest poems, flower culture, misused words, birds, and more. The May 1936 program was a debate: "Resolved that movies are detrimental to the young generation." The August 1936 program was dedicated to coping with the heat. Roll call was "Hot Weather Drinks," followed by "Suggestions for Hot Weather Lunches," and Mrs. Rogler offered "Ways of Keeping Cool."

The June roll call in 1929 was "The Disease I Fear Most." That was eleven years after the great flu epidemic. Children were still dying in those days of diphtheria, whooping cough, scarlet fever, pneumonia. On August 20, the roll-call question was "What do you consider the

---

Excerpted and adapted from Wes Jackson, *Becoming Native to This Place* (Lexington: University Press of Kentucky, 1994).

most essential to good citizenship?" In September of that year, it was "Birds of Our Country." The program was on the mourning dove.

What became of it all?

From 1923 to 1930, the program covers were beautiful, done at a print shop. From 1930 until 1937, the effects of the Depression are apparent—programs were either typed or mimeographed and had no cover. The programs for two years were missing. In 1940, the covers reappeared, this time typed on construction paper. The print-shop printing never came back.

The last program from the box dated from 1964. I don't know the last year Florence Johnson attended the club. I do know that Mrs. Johnson and her husband, Turk, celebrated their fiftieth wedding anniversary, for in the same box were some beautiful white fiftieth-anniversary napkins with golden bells and the names Florence and Turk between the years 1920 and 1970. A neighbor told me that Mrs. Johnson died in 1981. The high school closed in 1967. The lumberyard and hardware store closed about the same time, but no one knows for sure when. The last gas station went after that.

Back to those programs. The motto never changed. The sweet pea kept its standing. So did the pink and white club colors. The club collect—written by Mary Stewart in 1904 and popular in women's groups not only in the United States but around the world—persisted month after month, year after year. Here it is, "A Collect for Club Women":

> Keep us, O God, from pettiness; let us be large in thought, in word, in deed.
> Let us be done with fault-finding and leave off self-seeking.
> May we put away all pretense and meet each other face to face, without self-pity and without prejudice.
> May we never be hasty in judgment and always generous.
> Let us take time for all things; make us grow calm, serene, gentle.

Teach us to put into action our better impulses; straightforward
and unafraid.
Grant that we may realize it is the little things that create
differences; that in the big things of life we are as one.
And may we strive to touch and to know the great common
woman's heart of us all, and oh, Lord God, let us not forget to
be kind.

By modern standards, these people were poor. There was a kind of
naivete among these relatively unschooled women. Some of their po-
etry was not good. Some of their ideas about the way the world works
seem silly. Some of their club programs don't sound very interesting.
Some sound tedious. But their monthly agendas were filled with de-
cency, with efforts to learn about a wide variety of items: birds, our
government, and how to cope with their problems, the weather, and
diseases. Naive in some ways, perhaps, but they were living up to a far
broader spectrum of their potential than most of us do today.

I am not suggesting that we go back to 1923 or even to 1964. But I
will say that those women were farther along in the necessary journey
to live responsibly, even as they were losing ground, than we are.

The Matfield Green Women's New Century Club.

Uncle John

Uncle John's wife, Aunt Minnie, had died. Aunt Minnie was my mother's aunt. She and Uncle John had one child, a son named Luther, who had married only briefly and had also died. If my memory is correct, when Aunt Minnie died, Luther's wife got the farm, which she immediately sold. Uncle John had little money and no place to stay, so he came to live with my Aunt Ruth, my mother's younger sister. Uncle John had what he called a game leg, with an open sore that never healed. He had an upstairs bedroom in Aunt Ruth's big house and took his meals with her and two of her kids who were still at home, my cousins Danny and Martha.

Aunt Ruth lived on the farm next to ours, on the farmstead of my mother's parents. Aunt Ruth's husband, Uncle Art, had died, not in the war but during it. Their oldest son was fighting the Japanese in the Pacific.

I would go over there in the evenings to play board games and cards with Uncle John and whoever else was interested. Mostly it was Uncle John and me. When bedtime came, Uncle John would hobble up the steps, and Aunt Ruth would call after him, "Uncle John, did you apply the salve and dress your leg?" He would call back, "Thank you, Ruth dear, it is taken care of." He did not lie, for I had seen Uncle John expose his sore to the dog, who dutifully got up, walked over, and licked the wound. This, apparently, was all the treatment that Uncle John thought the leg needed. I have had people learned in the art of healing describe the merits of the dog treatment. To this day I have no idea what to make of it.

What brings Uncle John up is not the leg, or the dog, or the loss of his farm and his dear wife, Minnie, and his beloved son, Luther,

Excerpted and adapted from Wes Jackson, *Consulting the Genius of the Place: An Ecological Approach to a New Agriculture* (Berkeley, CA: Counterpoint Press, 2010).

though volumes of tears would have been shed in that kitchen by the cookstove (if it was winter) over those losses. What I want to talk about is the game of Monopoly, which the two of us frequently played.

I won every time, and not, I think, because I was a child and he let me. Every time, it would be the same story. We made the rounds as dictated by the roll of the dice. Uncle John might land on Park Place. He would look at the price, count his money, roll his cigar from one side of his mouth to the other, and always decline. "That's a little steep for me," he would say. But when Uncle John landed on a low-priced property, Baltic, for example, if he had the money he might snap it up.

You can appreciate where this story is headed. I ended up with the expensive properties. He ended up with the cheap ones. I put up houses and hotels every chance I got. He might sprinkle some houses on those cheap properties but would decline the opportunity more often than not. It wouldn't be long before he landed on one of my properties, often with one or more houses or a hotel. When he heard the rental price, it would hit him like a force. If he was low on cash, he would ask, "Can you hold off until I pass Go, Sharon?" (Remember that Sharon is my first name, which I went by until college.)

At first, I would hold off. But then he'd hit another one of my expensive properties, and I'd suggest that he sell a house or mortgage one of his properties.

"Naw, I ain't gonna mortgage," he would insist. "I saw what happened to my neighbors in the Depression. I vowed never to mortgage my place, and I kept it—me and Minnie. We just cut back." And then he would say, "I wish I had that place now. I could still make a go of it, even with my game leg, with a few chickens and my garden."

Well, that's how I beat him every time. I beat him because the game is rigged. I beat him because he was from the old school, which operated, to quote Milton, "according to the holy dictate of spare temperance."

My Life with Plants and Their Ecosystems: 1936–1952

1936

Strawberries, asparagus, rhubarb, blackberries, bluegrass pasture, potatoes, tomatoes, radishes, table beets, peas, onions, parsnips, sugar beets, carrots, sweet clover, oats, alfalfa, field corn, cucumbers, and popcorn. Ten head of livestock, including horses and mules and two milk cows, were on the pasture. Confined elsewhere, but eating plant products, were hogs and chickens. Farm records show all of this was well within five hundred yards of my first breath that June in 1936.

Those twenty crops had weed competitors that had to be destroyed using human and draft animals, all sun-powered labor. A human muscle–powered wheel hoe would destroy weeds between rows; a handheld hoe was used for those within rows. Fence-line weeds accommodated quail, but there were more than quail in that fencerow. Now and then a young pullet would steal out, go feral, build her nest, lay her eggs, and hatch and raise her chicks, which in turn became wild like their mother. What became of them? Chicken hawks maybe got 'em. Besides the homestead's yard trees, shrubs, and flowers, on adjacent property there was a woodlot with squirrels. A few trees were on Soldier Creek, which bounded our property on the north.

In 1936, the Dust Bowl was already famous, with the headline version farther west of us. The Soil Conservation Service was a year old. Roosevelt was president. Hitler was in Europe. The entire country was in the middle of the Great Depression. Numerous farmers out west had abandoned their farms, but not the valley farmers. Lots of people

Excerpted and adapted from "My Life with Plants & Their Ecosystems: 1936–1976," in *Big Botany: Conversations with the Plant World*, ed. Stephen H. Goddard (Lawrence, KS: Spencer Museum of Art, 2018).

were in the fields, along with crop and animal diversity unimaginable today. Diversity was nearly everywhere, with farmers growing food for Junction City, Manhattan, Wamego, St. Marys, Rossville, Silver Lake, Perry, Lawrence, Tonganoxie, Bonner Springs, Kansas City, and more. It was local and largely, but not completely, organic. Fried chicken in the summer, the fattening steer and butcher hog in the fall and winter. The rendering of lard and the making of lye soap for the weekly Monday-morning washday. People ate what was in season. Summer produce was canned.

"To write is to discover," Wendell Berry once said.[1] I have discovered as I write this that no matter what the task—field preparation, planting, cultivating, hoeing, harvest—I did not hear the word "botany" before college. We knew the common names of all the crops and some of the Latin names. And we saw an agricultural paradise, an abundance of plants for food, for flower gardens, for beauty. Weeds, never part of our botanical paradise, were competitors to be eliminated. Some were fiercely hated—crabgrass, bull thistle, bindweed, Johnsongrass.

Why was the term "botany" not used on the farm? Botany has been defined as "plant science." Agronomy scientists would show up at our farm "down from the college," Kansas State College, as I mentioned in the introduction. All were kindly men who respectfully offered advice to our family. I never heard them use the word "botany," but I don't recall them using the word "scientist," either.

The farm required countless hours of hoeing, often with a small crew. We sharpened our hoes in the morning and touched them up now and then, usually while resting at the end of a row, and at noon sharpened them again. To be efficient, there was a right way to properly file a hoe and only a few right ways to attack the weeds. A certain amount of character was on the line, and it didn't take long to get it right.

Nearly everyone was more or less, as we put it, in the same boat, and by that we meant economically.

1941: The War Years

In the early forties, the war was on our minds. I was old enough to remember, but I don't recall details of Pearl Harbor on December 7, 1941. My two older brothers were drafted, not to return until January 1, 1946. While they were gone, a steel-wheeled Ferguson tractor partially replaced the horses and mules. A gas-powered Planet Junior cultivator replaced the wheel hoe. Fossil fuel was now available to help kill weeds. Planting, hoeing, and harvest continued for a while after the war, but by the mid-1950s, our summer roadside market, formerly displaying our fruits and vegetables along Highway 24 and 40, three miles from Topeka, stood vacant. We drove our produce across the river. "Middlemen" bought the produce and delivered it to the various neighborhood grocers. But that, too, was on its way out.

1951: The Flood and the US Army Corps of Engineers

In "One Thing Leads to Another," I mention the '51 flood, which was higher than in 1903. Our farm was on the second bench, but even so the flood water ran a few inches deep across the floor of the barn and house. The crest came on Friday, July 13. When the river returned to its banks, there was a stench throughout the valley. Wheat straw draped the fences. Many livestock had been hauled to the hills, but many had been stranded and many drowned. Hill people took in valley people. The stories were pretty much the same up and down the valley. When the water went down, the agricultural world and the culture in the valley began to rapidly change.

Controversy about flood control broke out. Tempers flared. Large dams on the tributaries were proposed. As I mentioned in "One Thing Leads to Another," and it bears repeating, my dad thought the valley belonged to the river—an insightful observation, quietly stated. But my siblings, all older than I, were what was seen as "progressive" for that time. The US Army Corps of Engineers was ready. Soldier Creek behind our fields was straightened and levees built. No trees along the creek now. No more swinging out over the creek on grapevines. Dams were built on the major contributors to the Kaw. Towns that would be flooded on those tributaries were moved, as were the graveyards.

That memorable half decade after the war was over in a flash. The horses were nearly all gone. The hired hands were soldiers returned from Germany. I relentlessly tried to get details about their wounds and battles and would stop hoeing to question them. They kept hoeing. One told me what he and others had done to some Germans just captured. They told them to run. They did, and the Americans mowed them down. My brothers had been in the Pacific. They would not talk about the war to me.

It was clear that I talked too much. If you talk too much while hoeing, you miss weeds, get sloppy, especially on weeds within rows. You might not cut the root deep enough. You might use the point on the hoe too much. Better to stoop and pull the quack grass. Grasp it at the ground, wiggle and shake. When a brother drops back and comes behind you in your row to catch what you missed—well, you have been disciplined without a word spoken. Talk about the Germans or about other matters I thought of mutual interest stopped, at least for a while.

1952: A Pivot Point

In June 1952, less than a century after the Battle of Black Jack, which I mentioned in "One Thing Leads to Another," my sixteenth summer, I went to the prairies of South Dakota to work on a ranch belonging to a childless couple. Ina, an eccentric first cousin of my mother, and her equally eccentric Swedish immigrant husband, Andrew, were in their sixties and seventies by then and had somewhere between three and four thousand acres of prairie land.

They lived north of the town of White River. South was the Rosebud Reservation, home to Sicangu Sioux, and to the west the Pine Ridge Reservation, home to Oglala Sioux—two of the seven tribes of the Lakota Nation. That summer, I encountered the difficult reality faced by Native Americans as well as the differently difficult existence of the many immigrant ranchers. Mostly Scandinavian, they, too, as Wendell Berry put it, "came with visions, but not with sight"[2]—visions of former places but not the sight to understand where they were. To their credit, they learned in a hurry that most of that area was best used for bovine grazing.

On Saturdays, Native Americans came to town in buckboards pulled by teams or, more often, in pickup trucks. Their bovine grazer, the bison, was gone. I was mostly ignorant of where I was in historical time, nor did I know the word "worldview" or understand the clash of worldviews in front of me. I wish I had been more aware because on those dusty streets of White River, I must have exchanged glances with older Indians who, as children or young adults, might have witnessed the massacre at Wounded Knee in 1890. Some of them must have had relatives at the Battle of the Little Big Horn in 1876.

The major gift to me was to live in a land of mostly native prairie. It was there that I experienced my first intimate engagement with a landscape featuring a vegetative structure determined more by nature's ecology than by a people's culture. Even though the bison had

The author at age sixteen, the summer I worked on the Swan Ranch in South Dakota. This picture was taken at Valentine, Nebraska, fifty miles south of White River.

been replaced some eighty years earlier by their domestic bovine relatives from Europe, the land was less disturbed by human tinkering than most landscapes appropriated for human food.

Ina was Andrew's second wife; his first had been Ina's sister Bertha. Andrew and Bertha had homesteaded one half section and Ina another. When Bertha died after a twenty-year marriage, Andrew Swan and Ina Stover married, joined their holdings, and continued to add land. On some Sundays, I rode horses over those prairies with two teenaged brothers whose father was known at the time as a "half-breed," a term now viewed as derogatory. They told me of wildly unfamiliar adventures—both their own (they had roped a deer) and their parents' and grandparents'. They pointed to a hill where their Native American grandfather had trapped eagles. How did he use the feathers? I didn't think to ask his grandsons.

That prairie landscape was mostly unplowed then and still is today. The horse, then central to life, is less so now.

Andrew Swan's second wife,
my mother's cousin Ina Stover Swan.

Out on Ina and Andrew's ranch, other than the moon and stars, the only lights were dim ones from the towns of Murdo and Okaton across the river, twelve to fifteen miles away. It was a summer of branding, castrating, and fence mending, of dens of rattlesnakes to discover and fear, and of pond bass to catch. Many evenings on the ranch, I'd drive out on the point, a flat high place, to shoot prairie dogs or to see the hundred head of Andrew's horses on the range or among the trees along the river. Andrew kept them because he contended that horse trading had made it possible for him to be so solidly positioned. Today, I think of the slack that Andrew and Ina enjoyed to be able to afford those mostly unbroken horses grazing the unbroken prairie.

The Missouri River is only fifty miles from White River. Little of Jefferson's vision of the yeoman farmer was ever possible there. On those prairies, the land determined what could be done if humans

Shack where I lived in the summer of 1952, on the ranch of Andrew and Ina Swan. This was my first psychological fixation on native prairie.

were to stay. Some tried to farm the upland flats, but most failed. I loved everything about that country—the Native Americans, the rodeos, the rattlesnakes (from a distance), the Danish and Swedish immigrants, some with heavy accents, but all delighted with their land holdings.

And so the botanical contrast. In the Kansas River Valley, hoeing was endless. The ensemble of our crops would change every year, but on the ranch I did not miss the work that was required for the foods enjoyed on the farm: watermelons, sweet potatoes, cantaloupes, strawberries, peonies, sweet corn, potatoes, tomatoes, rhubarb, asparagus, and more. All required hoeing. From that, it was always a relief to put up alfalfa hay, harvest wheat, or dig potatoes.

The contrast between that truck, grain, and hay farm and the South Dakota ranch could not have been sharper. We were alone on the ranch most of the week. In Kansas, at our roadside market and

My Rural Life

when people came out from town to pick strawberries, our family met and talked with many people, and not just those from Topeka and the surrounding area but also travelers. None of that socializing compared to the life of the grassland, which supported Native Americans, ranchers, and rodeos. And so during my youth were two experiences with land. One, which became known later to me as the Jeffersonian agrarian ideal, was where culture dictated that ground be plowed, worked, and planted. The other, rangeland life, was where the plow had no place and was even anathema. I preferred the grassland.

Earlier I mentioned my grandfather from Virginia, who arrived in Kansas in 1877. With a small inheritance, he threw himself out onto the Flint Hills west of Topeka to run cattle on more or less free grass. At the end of ten years, he had enough money to go in with a partner and purchase 160 acres of sandy loam in the Kansas River Valley, where I was born.

I have often wondered why that grandfather purchased the farm, given that the grass had been so good to him. Why give up that way of life? It has taken me decades to acknowledge the power of culture and regional history, the power of a worldview, that term I first heard in college. My grandfather was born and raised in the Shenandoah Valley of Virginia. A Virginian! An agrarian! The ideal of the family farm as the source of virtue for the yeoman farmer had a history long before Jefferson championed it. Nevertheless, when thinking about the agrarian ideal, I could not resist looking at my road atlas and finding that my grandfather had grown up less than fifty miles as the crow flies from Jefferson's Monticello. Because of the history of that worldview, going back at least to the Greeks and Hebrews, the meaning of that distance in space is far less significant. Jefferson praised the small farmer from a cultural influence likely beyond books.

Much later, I finally had the language to better understand and describe how on the South Dakota ranch, ecological determinism is the major factor—how the land, not humans, determined plant behavior.

On those South Dakota prairies, west of the Missouri, if the original vegetative structure is plowed, a steep reduction in carrying capacity will shortly follow. In the Kansas River Valley, historical determinism ruled, and Jefferson's ideal could be a reality, at least for a while. South Dakota requires grazing. The Jeffersonian ideal can exist longer in eastern Kansas because of soils and rainfall, a more forgiving landscape. Across this continent ranges an ecological mosaic between and beyond these two places.

How Knowing the FFA Creed Came in Handy

Back in 1950, as a freshman at Seaman Rural High School, which sat to the north of Topeka and outside the city limits, I joined the local chapter of the Future Farmers of America (FFA). Any boy who took the vocational agriculture courses became a member, which required that one must learn the FFA Creed. I was happy to do so, since I believed every word of it. Every fourteen-year-old farm boy of the time did, or should have. I was elected FFA vice president in my junior year and president in my senior year. Such a high office required the holder to not only know the creed but lead others in its recitation, ending with "Fellow members, join me in the salute to our flag."

The creed was part of who I was, or what I wanted to be. Little did I know that when I was a freshman in college, enrolled in an introductory English class, my memory of that creed would come in handy. Here my story begins.

I was in the first semester of my freshman year at Kansas Wesleyan University at Salina, where I had gone with no higher aspiration than to play football and run track. KWU required that all students take the yearlong English course, which was divided between literature and communication. Dr. Johnson, the literature teacher, caused us to confront realities that few of us farm or small-town rural kids had known so far. From the boys' dorm, I remember such utterances as "What did you think of that goddam Beowulf?" "What was that Nibelungenlied all about? That wasn't English." "That Chaucer is something! Old Chanticleer and Pertelote, that 'Nun's Priest's Tale.'" "How about that 'Miller's Tale' and that gal that hung her ass out the window for the guy to kiss?" I was surprised we were allowed to read such stuff, given that KWU is a Methodist college. The English part was hard; Dr. Johnson was tough.

But now about the FFA Creed and how it came in handy. Mrs. Carlisle, the communication teacher, was steeped in Shakespeare and

Author as an eighteen-year-old freshman at Kansas Wesleyan University, 1954.

had directed many of the plays. In class she let us know that over the course of the semester, each of us would be called on to give an impromptu speech of three minutes or so. She placed our names in a jar and pulled out one or more each period. I guess this was to teach us to be ready to stand on our own two feet and say something when called on—to get us toned up for some real-world experience. Well, I had given the assignment little thought. I was busy. There was football practice and my job at the Pennant Café, where I washed dishes and pots and pans two hours a day for my two meals. I had an athletic scholarship, and I had to honor that. Then there was the girlfriend back home I had to write to as I looked at her picture on my desk. That romance didn't last until Christmas. Guess I was too busy otherwise. You know what I mean.

I had forgotten to prepare if called upon at any time to stand before the entire group, twenty-some students plus Mrs. Carlisle. As bad luck would have it, my name was drawn first. Stunned I was, but I rose from my seat, careful to see that the top of my desk was in order, and moved to the aisle past a few students to where I had a clear shot to the front. Seeing no impediments, I made the journey to where I was expected to perform. I straightened my bearing as I walked to the front, where I turned, faced the class, pushed my chin to the left and right, and gave a commanding look to the crowd, still without knowing what to say. But as you will see, the Good Lord delivered me. From what source I know not. Methodists don't speak in tongues. Many Methodists will pray out loud in public, though Jesus told them to "enter into thy closet, and when thou hast shut thy door, pray to thy Father which is in secret; and thy Father which seeth in secret himself shall reward thee openly" (Matthew 6:6). We are supposed to be in an attitude of prayer at all times. We are to "pray without ceasing." I reckon that is what I must have been doing as I made my way to the front.

But the first words out of my mouth were "I believe in the future of farming, with a faith born not of words but of deeds." I was off and running and delivered the entire FFA Creed without effort, concluding with a heartfelt passion, "I believe that to live and work on a good farm is pleasant as well as challenging, for I know the joys and discomforts of farm life and hold an inborn fondness for those associations which, even in hours of discouragement, I cannot deny." I did not call for a salute to our flag. Instead, I paused. I studied the crowd and nodded to Mrs. Carlisle. I walked back to my seat as my buddies whispered such utterances as "You sonabitch! The gawdamn FFA Creed? Holy shit!"

But what did I care? Mrs. Carlisle was swooning. I was lucky that she appeared not to know the FFA Creed and hence was not aware that my oration was not original. She said, in a voice that let me and others know what a moving, eloquent speaker I was, "Oh, Mr. Jackson! You have a future as a speaker."

Well, the guys all got over it. It was only gradually that I became ashamed of what I had done. Plagiarism is about the worst offense for a speaker or a writer. Where did my shame begin? I can't say. Perhaps it came to me on campus when Brother Cassell, professor of religion and Bible, greeted me with "Good morning, Wesley. What is the condition of your soul this morning?" Well! I had to think about that.

W. E. Cassell was professor of religion and bible at Kansas Wesleyan University. When I asked him if he expected to go and be with Jesus when he died, his response was, "Wesley, I never like your questions, but no. No, but I do believe that values are eternal." A thorough-going scholar. Photo courtesy of Terry Evans.

Brother Harley at the European Union Parliament

The cast for this one-act play is my wife, Joan; my brother Harley, eight years older than I; a German guide; and me. The setting is the Parliament of the European Union in Brussels. The year is 2000.

I have given a short, informal talk earlier in the day to the European Greens. Now they're gone, and the German guide, under the direction of a nonprofit in Stockholm, resumes his obligation to show us around. All is going fine until the guide says that, given the additional countries that will be coming into the union, this parliament building will not do. They will have to build another one.

With that sentence, brother Harley leaves the three of us and walks to the back of the assembly hall. The German registers his departure with a glance and continues his narration to Joan and me. We listen with more than dutiful interest. I'm thinking Harley should be here listening with comparable interest, but he continues his walk to the back, looking left and right, up and down. During a short pause in the guide's narration, I receive an order from the back with a tone and urgency that I might have heard back in the fields or at the barn on the farm. I see his long arm attached to a six-foot-three frame, with that index finger beckoning me over. I immediately become eight years old and he sixteen. I dutifully make my way to the back of the auditorium.

This time he wants my opinion on what he's pondering. I've pretty much already figured out what's on his mind, and so I'm not surprised when he asks, while thumping with the back of his hand against a wall, "You think this is a bearing wall?" The German guide and Joan arrive from the front while I study whether this is, in fact, a load-bearing wall (that is, a wall that doesn't just divide a room but also supports the weight of a floor or roof above; if it is a bearing wall, one has to be careful about knocking it out). Harley asks the German the same question—no matter that he has already determined that it isn't, and even if it were, it could be dealt with using steel beams. He consults the German again: "This space at the back—what do they do with it?" The German doesn't know. He says something about receptions. Harley says nothing, but I know he must be thinking that receptions are frivolous, maybe even thinking they could bring their own lunch or eat ahead of time.

Harley's questioning isn't over. "How many did you say would be added?" The German gives a ballpark approximation. Harley then begins counting seats and rows to get a rough estimate of this broad-at-the-back/pinching-in-at-the-front chamber, again sizes up the non-bearing wall, surveys the additional space available if that wall were removed, and declares, "I don't think you need a new building. Look here. All this room here at the back."

Harley didn't follow up. Maybe he thought it was enough for the German to hear and then get back to whoever might have a say in such matters. Harley was born in 1928, when the exercise of thrift and frugality were common. No need to build new if the existing structure will do the trick.

Schooling, Formal and Informal

*We often learn the most from the most unusual
people, in places we weren't looking for,
in ways we didn't expect.*

My Start in Botany: Earning a C, the Hard Way

In the fall of my junior year at Kansas Wesleyan University, I began to take upper-division courses as a biology major. One of them was General Botany. Getting lower than a C in any course in one's major meant that it did not count. Well, that semester I earned a D in that botany course. It was not an auspicious beginning for my academic career. I had no bona fide excuse. I deserved a D. Yes, I was working part-time. Yes, I had a girlfriend. Yes, I was playing football. I was missing some labs, but I was not missing practice. I was enjoying football, and if credit had been offered, I would have received a passing grade there. Perhaps football went well since the exams were weekly and public.

After a D in your major, what can a poor soul do? Contrition and humility before Professor Robinson were not going to hack it. Typical of Kansas Wesleyan professors, he wanted fifteen weeks of work, not mere repentance of sins. I had not measured up. When I told him that a D was not allowed in any major field, he said something like "OK, I'll give you six weeks and give you another exam, and if you get an A on that exam, you get a C for the course." I accepted the deal, got an A on the exam, and salvaged a C for the course. I went on to play on another championship football team in my senior year, but I also paid more attention to my courses.

Toward the end of my senior year, I failed to get any job offers to coach or teach high school biology. There was nothing left for me to do but try for graduate school in botany at the University of Kansas (KU). Lucky for me, Professor Robinson wrote what I was later told was a glowing letter to the KU department head. I was admitted and

Excerpted and adapted from "My Life with Plants & Their Ecosystems: 1936–1976," in *Big Botany: Conversations with the Plant World*, ed. Stephen H. Goddard (Lawrence, KS: Spencer Museum of Art, 2018).

had an assistantship to teach biology and botany labs over the next two years while working toward my master's degree.

So now I was at KU as a beginning student in the botany department. These were two years that became among the most important of my life. The courses, field trips, research, and thesis brought a kind of satisfaction and plain joy I had not imagined.

I started out in plant taxonomy, and I did want to know the plants. In taxonomy you are expected to learn how they got their names. Fair enough. Humans have always recognized "kind." Why should there be any problems in doing so? There are rules—in fact, international rules—for botanical nomenclature. But rules are administered by people, and botany seems to have more than its share of weird and wonderful people, especially among the taxonomists. After getting to know some of them and hearing and reading about their disagreements, I believe that as a group they are beyond ordinary eccentricity, whatever that is. One famous botanist once said that taxonomists are like mice hiding behind herbarium cabinets hating one another.

When it came to classifying the various plants into genera or species or subspecies or varieties, some taxonomists are "lumpers" and some are "splitters." Lumpers focus on signature similarities, while splitters are more attuned to differences. But all are very serious and seemingly opinionated about their concept of order, no matter what particular species or genus they are working on. As eccentric, weird, wonderful, and entertaining as they were, I felt that I would not make the cut with all of those legalistic details and varying interpretations, and frankly, I lost interest.

I was interested, however, in genetics and evolution. I had received an A in genetics from the same professor who had awarded me the D in botany. Lucky for me, there was a new hire in botany at KU, a young PhD named R. C. Jackson, no relation of mine. He had discovered the species with the lowest chromosome number in plants, N=2. The plant was *Haplopappus gracilis.* I wanted to work with him in the general area of cytogenetics. After several candidate species were brought to my attention, I selected the genus *Ratibida* in the sunflower family.

Summer was spent collecting plants in the Southwest, traveling up and down and across Mexico with fellow graduate students in a KU van, sleeping out, and preparing meals with food from small grocery stores and private stands. After supper we pressed our plants, changed blotters, tidied up our notes on what plants were associated with our collections. It was a great summer.

The next summer, again collecting for KU, John Morris, a graduate student from Wales, and I collected specimens all over western Kansas, sleeping in state and county parks, pressing plants, and changing blotters in park toilets to get out of the wind. Other campers and local citizens were curious about what we were up to and why. We explained as best we could the practical reasons to be collecting "weeds." Luckily, I was ready for this because the summer before, several of us had joined the nationally known taxonomist and department head, Professor Ronald McGregor, on a field trip, and a local rancher-farmer at a diner in Kanorado, Kansas, near the Colorado line, had asked what we were doing. Dr. McGregor said, "Looking to see if there are any new weeds coming into Kansas." The rancher-farmer seemed satisfied.

After earning my master's in botany at KU, I spent two years teaching high school biology, two years back at Kansas Wesleyan to teach biology courses, two and a half years working on a genetics degree at North Carolina State, four and a half more years of biology teaching at Kansas Wesleyan, and three years in the Environmental Studies Department at California State University in Sacramento. Then came The Land Institute.

Professor Robinson would have been within his rights to let the D stand. Where would I be today if he had sent me away?

I've talked to several successful professionals, long in their years, who acknowledged how little any conscious planning had to do with how their lives turned out. In such conversations, the word "luck" was mentioned more than once. So I'm not alone in this feeling. Wendell Berry once told me told me that his father, reflecting on a life full of meaningful work and loving family, confided, "I had nothing to do with it." So it has been for me.

Dr. Wassermann

A few years after World War II, Dr. Felix Wasserman would go to Europe every summer—take Icelandic Airlines, buy an E-Rail pass, sleep on the train most nights as he went from city to city, visiting art museums, libraries, maybe old friends. He was, I suppose, what at one time people called a "bohemian." Such folks were not the norm in Salina. I certainly benefited from his time as a professor at Kansas Wesleyan University.

Dr. Wasserman had fled Hitler's Germany in time to make it to New York. His wife, who had a PhD in art history, was soon to follow. But she never made it, and he never knew what happened. Both were Jews.

Living in New York for a year, he washed dishes in exchange for meals and ten dollars a week and slept in a flophouse. He went to Mississippi, taught at a small college, and met a young Mississippi woman named Christine. They married and had one child, a boy they named Felix Ludwig. They called him Puss.

He didn't stay long in Mississippi. In the early 1950s, he was offered a job at Kansas Wesleyan to teach languages and the humanities. He knew seven languages. He taught Latin, German, French, and Spanish plus various courses in the humanities. He was my French teacher, and I was a terrible student of French. I might have complained because it was French with a heavy German accent, but that was not it. I was actually better in German. Anglo-Saxon for me.

I liked his humanities classes. I especially liked the medieval European art assignments, where we were to find paintings we liked and write an essay. I first selected Van Gogh's *Potato Eaters*, maybe because I liked potatoes, had grown them, picked bugs off them, dug them, and sorted them. The people in that painting are hungry. Clearly, they need more potatoes, and I wondered what that was about. Maybe not enough of them, or not enough green leafy vegetables. I learned that this was an early painting by the artist. I could see that he knew how to paint details. Had I first seen his later, post-Impressionist paintings, I would have suspected that he did not know how to paint at all. But that *Potato Eaters* showed me that he had the stuff, that he was good. Dr. Wassermann was patient with my critique, and I learned a little about the Impressionists.

I became increasingly interested in Van Gogh, read the 1934 biographical novel *Lust for Life*,[1] and was intrigued by the part where he cut off his own ear. Many years later, I learned that he may have suffered from lead poisoning. He used lead-based paint and may have ingested enough lead to perhaps cause both his insanity and his post-Impressionist period. The Roman nobility who could afford advanced plumbing often had lead in their pipes. They suffered from it, too. It seems that many chemicals were around and used to human detriment long before the industrial chemical period. Men who made hats often went mad; "mad as a hatter" was a common term. I understand they used mercury in the shaping of hats. Felt hats, popular at the time, came from animal skins dipped in mercuric nitrate to separate the fur from the pelt and mat it together. Then, the felt could be shaped into a hat. Maybe it was both lead and mercury?

I would have liked to talk to Dr. Wassermann about how much fine art may have been due to mind-altering chemicals. We could have had a conversation about some art being due to the limits imposed by some chemicals. Actually, now that I think of it, I would not have wanted to follow up on that with him. He would have wanted me to look into it, write something up, let him critique it. I preferred to remain a dilettante and avoid the risk of being a charlatan, which is far worse. Being a hopeless dilettante has its merits, but not for Wassermann.

Now for the near-true stories I am about to tell. There are still a few alive who knew Dr. Wasserman and can back me up on the essence of my telling.

I owe him a great debt. Beyond smatterings of ancient history and medieval art, he introduced me to Alexander von Humboldt, the great naturalist and explorer of the late eighteenth and early nineteenth centuries. Dr. Wasserman had published scholarly papers about him. And while he may not have been the most eccentric person I have known, he is a contender for the title. His passionate, eccentric, and otherwise weird and wonderful ways do lay some of the background for the story.

Dr. Wassermann seemed unaware of what was going on with the external part of his whole body and what was attached, meaning his shoes or clothing. However, now and then there must have been some discomfort, for he would make corrections we would all notice. When he was lecturing, his glasses would come down on his nose, partly because his head was often tilted upward toward the ceiling or where the wall and ceiling meet. He would push his glasses back and discover that he still could not adequately see. He would then inspect them and discover what we all knew—they were dirty. Without bothering to remove them, he would take his thumb and forefinger and rub both lenses, only to discover that he had only managed to redistribute the smear. With his glasses still on, he would sometimes pull out his wadded-up handkerchief to clean them. I doubt that anyone believed that handkerchief, no matter how it was applied, would do any good for his vision problem. He would return his wrinkled handkerchief to his front pocket, having given up.

A few times he appeared in early-morning class wearing two different shoes—one a mud grip, the other a smoothie. Someone would point this out to him, and he was surprised every time. Regardless, he kept the same shoes on all day for all classes.

One time the stitching on one of the soles had broken, and the sole was loose from the tip back toward the front of the heel. When he walked, he was forced to lift his leg and throw the foot out in front so as not to have the sole double up under his foot. He was sufficiently successful in this task that as he walked around the room, lecturing away, chin up, looking at the ceiling, glasses down as he was telling us something out of Thucydides or some other Greek or Roman historian, I couldn't help thinking of the German goose-step as he kicked the leg with the flapping sole.

The fashion people would have had a field day with him. He could have been one person's full-time job to meet the standards of the time. Maybe he just needed a full-time butler.

Now for the best part. One day the whole class discovered that Dr. Wassermann wore striped undershorts. He was lecturing away, looking at the ceiling and thinking that he was pulling up his trousers. He was actually pulling up his shorts. As he walked and talked, he pulled and tugged on both sides with a firm grip, using both hands. This went on until it reached a point at which he discovered that he had pulled everything up, thereby creating a "discomfort zone."

This was a discovery that required correction, sure enough, but he never looked down to analyze the problem and discover that it was his striped shorts that had been pulled up to you-know-where. His instinct was strong when it came to crotch management. Still lecturing, he needed no visual inspection. He just raised one knee and kicked outward, then did the same with the other leg to reposition what was bothering him. How long this went on, this kicking with one leg and then the other, now and then grabbing the waistband of his trousers, I can't remember. The whole scene is better demonstrated than told.

I am sure the reader will think I have conjured up some of these details to make the story better than reality. But these "wardrobe malfunctions" were not rare. As we students settled into our seats one morning that I remember clearly, with a quick inspection we realized that another show was about to begin. We boys—OK, we young men—laid down our writing instruments. The young women and the preministerial students were more or less ready to ignore the coming show and to pay attention instead to the scholarly outpourings of con-

Felix Wasserman at his desk, 1973–1974. He was an art history scholar and fluent in seven languages. Unlike anyone I had ever known or imagined. Courtesy of the Department of Special Collections and University Archives, Raynor Memorial Libraries, Marquette University.

tent sure to come forth without interruption. There was the flapping shoe again and the striped shorts already above the pants line. At least the shoes matched this time. Maybe that wasn't unusual, but we did have the flapping sole, colorful shorts, and kicking legs as his crotch management was once again launched.

Dr. Felix Wassermann, Christine, and Puss are long gone. In spite of my poor performance in French and, by his standards, my poor performance in the introduction to the humanities course, he was one of the top-notch teachers I was lucky to have. All of his physical antics in the classroom and elsewhere are not what I go back to the most often, more than sixty years later. It's not the quirkiness but the intellectual depth of what is humanly possible that I cherish.

Doc Horr

Worthie Horr—that was his name—grew up on a farm in northeast Kansas and went to KU to learn enough botany to start and run a greenhouse. While at KU, he became interested in plant physiology, graduated, and went on to the University of Chicago, where he earned a doctorate. He returned to KU, where he stayed until he retired from the Botany Department as professor emeritus.

His broad interests and much of his expertise were due in part to military service during World War I as well as the Great Depression. When he was a professor during the 1930s, KU was short of both funds and faculty, which required him to teach several courses. Even without the diverse teaching load, he still would have been a consummate naturalist and a down-to-earth good person. Like many of the old botanists of that era, he was quirky at times, asking such questions as "When the hell does a tree die?" One of his final exams consisted of this statement written on the blackboard: "Discuss the colloidal properties of protoplasm."

There was no air conditioning on campus in the fifties, and Kansas weather can get hot. Doc Horr kept his office door open to make himself available to anyone who stopped by. When it got too hot, he would take off his shirt, exposing his undershirt. I don't know how many undershirts he had, but if my memory is correct, none of them was without holes. Here is a typical vision of Doc in his office: door open, seated at his desk, sometimes with his feet up, reading a paper or professional journal, often with cigarette in mouth. In those days, people smoked more or less when and wherever they wanted.

He had been gassed in France during World War I. As a result, part of one side of his face was paralyzed, and he had lost about one-third of his lung tissue. When I asked him how that happened, he explained that he heard the gas shells, which on impact had a different sound than other munitions. He was a cook in the army, behind the front lines, but his deep curiosity got the best of him, and he decided to go to the front and have a look. One landed near him. That's a powerful curiosity that can draw someone toward fighting to learn about differences in the sounds that artillery shells make.

When he smoked, his cigarette would get wet. He would inspect it, notice the moisture, and, using his thumb and forefinger, break off the wet end and toss it toward the wastebasket near the door. "Toss it toward," is the key phrase; more wet ends seem to land outside than inside the wastebasket, leaving a trail to record his accuracy.

If I had a question for this wise scholar, I'd better have it firmly in mind if I expected to avoid being distracted by his holey undershirt, feet on desk, wet cigarette in mouth, butts around the wastebasket.

In his ecology class, he was particularly rich with eccentricities. He would frequently bring out his glass-lantern slides (that was an old technology even then, tracing its lineage back to Dutch scientist Christiaan Huygens in the seventeenth century). Most had a crack somewhere. They were all black-and-white. Several were pictures of the Indiana Sand Dunes, which he had studied while at Chicago, and others were from the prairies, taken by two great prairie ecologists at the University of Nebraska, J. E. Weaver and F. E. Clements.

He once took our class to what is called the Muscotah Marsh in northeast Kansas. This ecosystem was left over from the Kansan ice sheet. Countless soil and pool samples had been taken in and around the marsh area by various professor types, thus allowing scientists and their students to see, based on pollen samples, what the vegetation had been from time to time. Pollen anatomy became widely used during Doc Horr's lifetime to get a handle on the vegetation and changes over time. There were thousands of invertebrate shells, and if I remember correctly, the retreat of the ice had left behind living snails not found anywhere else in Kansas. I wonder if they are still there?

One summer I was his assistant in a summer botany course for teachers. Outside the classroom, one fall he paid me to help him park cars on his lawn across the street from the stadium during football games. (I have a hard time imagining a contemporary faculty member at a big school running such a side business.) He had built his house, using thick native limestone. The vertical part of the footing was some four feet thick in places, the walls maybe even two feet thick. It was stone all the way up, ending with two peaks, one with about a 12/12 pitch and the other more like 17/12 (for those who don't know roofs, that's steep). The first number in those ratios represents the height of the roof, the second the length of the base.

I helped him shingle his house one summer, and while we were shingling away, he on one side of the peak, I on the other, out of the blue came this: "When we went into France, we clipped the vocal cords on the mules." There is a sentence that cries out for explanation. I made my way up the steep ridge, looked over, and asked, "How's that?" Without looking up, hammer in hand, he simply replied, "Oh, hell, when we went into France, we had to clip the vocal cords on the mules. You could hear them braying ten miles off." That was no doubt an exaggeration, but I didn't question it, and that ended the exchange.

I have often thought of that moment and of how fortunate I was to be shingling with him that summer. It was nearly impossible not to learn something when around him. His range of engagement with the world would be hard to match: a complete naturalist, gassed in World War I, fought the Germans on a European battlefield where draft animals were counted as necessary in the "war to end all wars."

And I want to repeat what I have already said to emphasize it: Here was a professor trained in the specialty of plant physiology who knew the flora of Kansas partly because of the Great Depression forcing him, like others in the department, to also teach plant taxonomy, general botany, and ecology and be curator of the herbarium. He taught both undergraduates and graduates and directed graduate students. And, as suggested earlier, he could tell anyone interested about the flora of the area during all interglacial stages of the Pleistocene, based on pollen deposits.

Was he a saint? Of course not. He had many derogatory stories about past chancellors, past and present faculty, a department head who made his life difficult. Did he have a sense of humor? Yes, a quiet one. He once let us in on a fictitious secret, how he and Professor Robert Beer, the head of entomology, planned to open a business on the main street in Lawrence. It would be called the Beer and Horr House. There was never a mention of what they would be selling.

He was a Republican. Conservative or liberal? He did not like Franklin D. Roosevelt, especially some of the employment programs of the 1930s, which included planting shelter belts on the Great Plains during the Dust Bowl era. He thought that effort ill-conceived and set out to study those shelter belts. He had numerous shelter-belt maps from western Kansas counties rolled into tube form. Some he had made, some were from the government. He had written notes on what species had been planted when, how much precipitation the area received, weather conditions at the time, from what distance the trees had an influence on crop yields, how far from the belt the wind scooped down to create a trench or drift. The length and number of rows of trees were recorded, what trees died, when, and under what conditions. He could tell you when various tree species "cut their leaves" during Dust Bowl drought conditions.

After he retired, I would stop by on occasion to enjoy a visit and see how he was. His wife was not well. As a young mother, she had suffered a stroke that placed limits on her and added to his burden. Their only child, David, graduated from KU and Harvard, where he studied orangutans, got a PhD, reversed his middle and last names, and retired following a productive career as a biologist and educator.

Over the years in that period, I met several old botanists around the country. It seemed that they all had some eccentricity. Doc was one of them, for sure. One story about Doc was told by Professor

Ronald McGregor, the department head in my time. He had once been Doc's student and had taken over running the department's herbarium from him. The story goes that Doc was on a tall ladder in the storeroom, removing Pyrex tubes one by one from a box at the top of a tall cabinet. He would hold out one tube at a time and drop it. How long this went on, I can't say, but broken glass lay on the floor when the department head, Professor A. J. Mix, walked in, looked around, observed the scene, and asked, "What are you doing, Worthie?"

"I'm looking for Pyrex tubes," he replied.

Dr. Mix reached down, picked up a piece, inspected it, and said, "This says Pyrex," to which Doc Horr replied, "Thought the damned things weren't supposed to break?" Somehow he had missed that they were designed not to break when heated, not when dropped from seven feet or so.

This story caused me to think about the evolution of laboratory supplies and equipment and the unacknowledged limits to research. Once upon a time it was not possible to heat glass tubes without breaking them. So here is a question: What did scientists do before Pyrex tubes were common for laboratories anywhere? Laboratory science had been going on for three centuries. This was a new thing, this heat-resistant glass tubing. And another question: What made him think these tubes would not break? Had he seen an ad, or had a supplier stopped by and taken an order for this new glassware? Perhaps what he heard was just enough for him or for anybody to have heard of the Pyrex tubes and to believe they were not supposed to break *at all.* What was missing in his mind was the qualifier—under heat or open flame.

In our culture, it doesn't take long for a new innovation to go from a marvel to a ho-hum expectation. Pyrex is a heat-resistant glass, familiar in most laboratories. But for Doc Horr, all that was brand-new.

Back to the history of heat-resistant glass. Let's see where it goes. According to archaeologists, humans have been making glass for more than five thousand years.[2] Until the 1880s, glass was prone to break when heat was applied. Otto Schott, a German chemist and a glassmaker's son, discovered that adding boron to glass allowed it to be heated without breaking. What is called borosilicate glass was limited to specialty products for the next three decades. Then in 1914, physicist Jesse Littleton was evaluating borosilicate glass for use in railroad lanterns and battery jars. His wife, Bessie, wondered if the glass might be good for baking, so he sawed off part of a battery jar and took it home. Bessie baked a cake in it, and her husband's employer, Corning Glass Works, developed the first-ever consumer cooking products made of glass. They called it Pyrex.[3]

So now we are at 1914. Glass had the virtue over cast iron and earthenware of not imparting smells or tastes to food, but it was another twenty years before it really caught on, which takes us to 1934, a half century after chemist Otto Schott added boron. What was the holdup? Pyrex was expensive until production became fully automated.

Pyrex flasks, beakers, and test tubes provided a giant leap for laboratory science. No wonder the chemical industry was able to take off after World War II, and no wonder that a still-young plant physiologist born in 1890—in the midst of all this technological change—could assume that "the damned things were not supposed to break." Could Doc Horr not be forgiven for this small lack of knowledge? After all, he did know how to clip the vocal cords on the mules when trench warfare was common. He knew what gas warfare was like. He built his own house and shingled his own roof, parked cars on his lawn on football game days. He thought it was important that we think about the colloidal properties of protoplasm, and he watched with interest a citizen campaign to make Lawrence the Redbud Capital of the World. And he knew that shelter belts would suck moisture two to three rods away from a wheat field.

And now to wrap this up. With little urging from me, he would talk about his studies of shelter belts from time to time. I had been reluctant to ask Doc about the eventual fate of his maps. But then came an opening. Near our last visit, he offered me his choice books from his large library. I was moved and thanked him for the books. It seemed the time to ask him, "What about those maps, Doc?" And then the reply: "Oh, hell, did you want them? I wish you had been here last week. I burned them."

It was, for me, a tragic moment, and both of us sat in silence. I did not ask why. There was nothing to be said that would not resurrect, what was it—grief, disappointment, lack of a successor, abandoned scientific interest? After a while, we went on talking for a bit, and then I left.

A more cheerful final story, and a more fitting ending, has to do with an account he gave me at least twice. It has to do with his second and last trip to Europe—not to a world of gas warfare and trenches but this time to Vienna to visit his son, David. He attended a concert at the Vienna Volksoper, the Vienna People's Opera. He described that night as a moment in which he had never had so much pleasure. I did not ask him about the performance, but I have fantasized that in the concert hall that evening, the orchestra's last piece was Beethoven's "Ode to Joy." That would have been fitting for a man who most deserved it.

Harry Mason

> There is no doubt that the two generations between
> the late 1860s and the beginning of WWI remain the
> greatest technical watershed in human history.
> —*Vaclav Smil,* Creating the Twentieth Century

> In the old, familiar story of westward expansion failure is
> a parenthetical event. If it is not Satan in Paradise Lost,
> then who wants to listen?
> —*Mary Beth LaDow, "Chinook, Montana, and
> the Myth of Progressive Adaptation"*

Many people think of the late twentieth century as the era of great advances in technology, but Smil points out the numerous changes during the period between the end of the Civil War and the start of World War I. Many people think of that post–Civil War period as the final phase of the successful US expansion across the continent, but LaDow points out that we ignore the failures in that project and the often profound implications. Smil suggests that our focus on the technology of more recent decades leads us to misunderstand how the modern world came to be. LaDow suggests that people will ponder humanity's fall from grace in grand literature such as *Paradise Lost* but ignore the countless small failures that constitute our history.

Both of those insights come to mind when I think of Harry Mason and the Great Plains of the United States. He appreciated working with the technology of that era, and he saw the failure as well.

Harry, born in 1908, was raised in WaKeeney, Kansas, which is near the 100th meridian, almost halfway between Denver and Kansas

---

Excerpted and adapted from Wes Jackson, foreword to Harry Morgan Mason, *Life on the Dry Line Working the Land, 1902–1944* (Golden, CO: Fulcrum Publishing, 1992).

City on the dry plains. Midgrass prairie gives way to shortgrass here or, coming from Denver, the other way around, to tallgrass prairie.

Harry's parents had failed as farmers and then bought a garage in town, which failed, too. They returned to farming and failed a third time. They were about to move into town in April 1944 when his sixty-nine-year-old father died, probably in the cab of his truck, probably drunk. As Harry put it, "Before Prohibition, before the garage venture, before the dust storms and the Depression, he might have taken a drink of whiskey when offered in a social setting, but he did not have a drinking problem." It's easy to think that his family history made Harry particularly attuned to a culture's failures.

The family struggled, but it was in that garage that the young Harry Mason learned to repair the broken promises of the Industrial Revolution as a young user of the products of "the greatest technical watershed in human history," as Smil put it. I met him years later and only recently came to appreciate that he felt he was alive at a historic moment. But before I came to that understanding, I had the feeling that his days as a mechanic were the happiest of his long life. It is touching to read in his memoir, *Life on the Dry Line Working the Land, 1902–1944,* how he felt as he would drive out into the country to fix a neighbor's tractor in the field, how gratifying it was for both the farmer and Harry to "get the thing going again." It didn't matter that the hours were long, the winters cold, and the summers hot and dry. Traditional virtues—patience, hard work, attention to detail, and a good deal of love—went into helping the owners of those first machines of the still-early era of industrialized agriculture. His intellectual career that followed was fulfilling, but university life was probably never as much fun as garage life for Harry.

He ended up going to college at nearby Fort Hays State, where he liked his psychology professor. Probably for no better reason, he studied experimental psychology. After earning a PhD, he taught in several universities before settling at Kansas Wesleyan University, where I was first his student and where I received a D, along with a few others. Later, after I returned as a faculty member after my own graduate education, we became colleagues.

I am grateful for his contribution to my life in both phases, but not everyone would look back as fondly on Harry. He was, as the saying goes, not everyone's "cup of tea." He was brutally frank, often alienating others, which partly explains his short stay at those other universities. Harry had very poor hearing, and no hearing aid seemed to do much good. My dominant memory of him in faculty meetings was his tendency to cup his hand behind an ear to help him hear, with head bowed, eyes closed. Once he had heard, as often as not he would mutter, "Bullshit," under his breath, sometimes followed by "More bullshit."

Harry was an acquired taste, and for whatever reason, I acquired it. I counted him as a friend and useful mentor. When I was building our house on the property that became The Land Institute, I valued his help, not just with the labor but with figuring out how to solve problems that arose. Given that I worked without the aid of an architect, blueprints, or much experience building a house, using a lot of scrap material, problems often arose. Harry had a keen eye for how to make things work, no doubt due to his early years as a mechanic in farm country, where making do with the tools and materials on hand was the norm.

Harry's contrarian nature rubbed almost everyone the wrong way, but it was also a source of humor. The English literature professor once asked him, "What do you like to read, Harry?" He was widely read, but he paused to ponder, then said, "Directions." He wasn't being completely facetious. He seemed to always want to know how things worked.

Sometimes his humor was more biting. Once asked what he thought of a writer, Harry said, "He's a literary type. The truth isn't in him." Once addressed as "Professor Mason" during an encounter with an administrator, he responded, "I'm not a professor. A professor doesn't care. I care." He could paint with a pretty broad brush. But even when he wasn't being fair, he wasn't completely off base. He was sensing the failure of the modern university as it was becoming increasingly specialized and professionalized and, to him, losing the spirit that should animate higher education.

He enjoyed fixing engines but left farm country as the economics of rural America made it harder for people to succeed. He also knew he couldn't deepen his education out there. That led him to higher education, where I imagined he hoped to find like-minded people. But in that expanding post–World War II university community, he found himself out of sync there as well.

Harry's responses could cut to the heart of a question. When I was back at Kansas Wesleyan as a teacher (I'm not about to describe myself as a professor, given Harry's scorn for the term), I was talking with him about being vexed by the free-will-versus-determinism question. "Oh, Wes," he said, "that's the oldest debate there is, and there's no answer to it. Just enjoy what the world has to offer." Another time, when I was pondering why people behaved in a certain way, Harry made a philosophical point in memorable language: "Oh, Wes, it's in the meat."

He could have said, "That behavior is a product of human evolution, and, while subject to marginal modification by individual experience and cultural differences, it's largely a product of our shared genetic predisposition." If he had put it that way, I doubt that nearly a half century later, when asked a similar question, I would be repeating that long explanation. But every now and then, when asked about some aspect of human behavior, I find myself saying, "It's in the meat."

When Harry was born, Kansas was five years less than a half century old. When he was a boy and young man, Civil War veterans came in and out of the shop in WaKeeney. He tried to be a Christian early on but finally concluded that when he died, as he put it, "it's like a light switch." Even so, time and again I would detect that he was a secularized Christian. He gave money to needy students and needy causes. He had a welder in his basement and built many of the accoutrements that adorned the house, often as comical as useful. His salary did not go up along with those of other professors. He loved photography and turned out many high-quality pictures. Serious students used his darkroom at will. He loved his pickup truck, which served as a camper for him and his wife, Isabel. Nothing about it was fancy.

But back to the intersection of technology and failure. All our advanced technology is counted by most people as success. The conversion of all that prairie into the so-called "breadbasket of America" is counted as success. But what of the people—both indigenous and failed farmers—who were destroyed in all that success? What about the ecological destruction required for all that grain? Could Milton find a way to work that into *Paradise Lost*?

Scientifically Speaking

I eventually learned that science isn't a set of facts, theories, or methods but rather is a way—though not the only way—of being in the world.

A Field Trip with Three Great Scientists

For many scientists, there must be "the most memorable field trip," one that sticks out from all the rest. Mine lasted three days in September 1985, near Comptche in Mendocino County, California, about four hours north of San Francisco. Hans Jenny was one of the world's leading soil scientists of our time and author of the classic text *The Soil Resource.*[1] His friend Arnold Schultz was a University of California, Berkeley professor who coined the term "ecosystemology."[2] Those two gentlemen led J. Stan Rowe (a Canadian ecologist whose work has greatly influenced my thinking) and me up and down the ecological staircase of Mendocino.

The terraces were created by the uplift of the land along the Pacific shore by tectonic plates that slide under the continental plate and push all above it upward. From there, the action of waves and the rise and fall of the ocean level shaped the rising landform, leaving wave-cut platforms. These terraces at Mendocino were built over half a million years.

There are five such terraces for this story. Beginning at the ocean's edge, the first terrace is about 100,000 years old, the second 200,000, the third 300,000, the fourth 400,000 and the fifth 500,000. Terrace one features grassland. On terrace two, we see lush strands of redwood and Douglas fir. Terrace three is a transition zone with some bishop pine coming in. The fourth and fifth terraces support only what is called a pygmy forest, a fairly rare ecosystem on the planet.

Before I started up that staircase with my colleagues, I was a firm believer that any natural ecosystem was sure to improve—meaning it would add topsoil, increase in stability, maybe diversity, and, if not improve, at least stay good indefinitely. By the time we headed

Excerpted and adapted from Wes Jackson, *Consulting the Genius of the Place: An Ecological Approach to a New Agriculture* (Berkeley, CA: Counterpoint Press, 2010).

back toward Berkeley in the car, these pillars of my ecological understanding had been shaken. I wanted the soil to conform to humanity's
needs. What we label "good" is that which sponsors abundance for us.
I realized I was imposing my human conception of what constitutes a
"good" ecosystem on nature, and nature wasn't cooperating.

My concerns grew over the next several weeks. About four months
after the field trip, there came a letter from Hans saying that he was
not aware that there was any reason to believe in a concept of steadily
improving ecosystems. He said that such a "sunshiny belief rests on
neglect to appreciate the soil as a dynamic—either improving or de-
grading—vital component of land ecosystems."

In that same letter, he expressed concern as to whether he and
Arnold had presented Stan and me with "sufficient physical evidence
that the decline in soil and vegetation from the redwood–Douglas fir
forest on the second terrace to the pygmy forest is a natural sequence."
Plant ecologists saw instead two ecosystems; the redwood-fir forest is
a climatic climax, which is to say a steady state that is due to the influence of climate. The pygmy forest is an edaphic climax, which means
it is a steady state that is due to the soils. In other words, ecologists
had designated two different worlds. Hans thought they did not realize "that the two ecosystems might be on the same time arrow, merely
separated by a long time interval." Here is the crowning upshot: Over
a longer time period, the younger, more robust vegetation will lose
nutrients due to such natural forces as wind and water, aided by grav-
ity. No matter how robust they are initially, they, too, will become
pygmy-like since the leaching is relentless.

Fundamentalism of any variety tends to die hard, and I realized I
had been harboring a kind of fundamentalist view that natural ecosystems always improve and move toward stability. Many soil scientists like to dig down and create soil pits in order to see the vertical
layering of time. On the fourth terrace, there was such a pit. Here
I could sympathize with the church officials who refused to look

through Galileo's telescope (even if the story is apocryphal). Now here I was, standing and staring at the evidence, still insisting that good farming can improve the soil.

Yes, Hans said, but "the extent depends on what kind of soil, virgin or depleted, the farmer begins with." He thought it would be difficult to improve a good virgin Iowa prairie soil by soil-management techniques, except perhaps by applying N, P, and K (nitrogen, phosphorus, and potassium—the main nutrients in fertilizers). And, of course, "improving" a soil by importing nutrients from elsewhere is robbing Peter to pay Paul; the place from which the nutrients come is now the poorer for the transfer. It happens frequently, and I now call that an "acceptable theft"—acceptable, that is, in the eyes of those doing the taking.

It was the beginning of an important lesson to me, and since then, I have burdened myself, students, and colleagues with the following question: Why should our commitment to "nature as measure"—our belief that we should look to nature as we work out our relationship to the earth—provide us with easy absolutes or simple slogans? At The Land Institute, we have chosen to make nature our standard against which to measure human agricultural practices, which we believe to be appropriate. But that choice does not guarantee that we will find the simple principles that we may yearn for.

Nature does not organize life to fit our ideas of what is beautiful or best. It is we, not nature, who are loaded with notions of good and bad. Most of us are inclined to think that a luxuriant redwood forest is "superior" to a pygmy forest, perhaps because we find it more beautiful or majestic. But not Hans, who, in thinking about the pygmy forest, insisted that "nature might call it a biological improvement, an adaptation of vegetation to a changing substrate."

Now came a big question: If a reduction of fertility occurs due to leaching, wind, and water erosion, even with vegetative cover, why are there not pygmy forests, pygmy prairies, pygmy whatever, worldwide?

The answer lies in the actions stemming from the bowels of the earth, where various geological events happen, such as mountain formation, volcanoes, mostly in geologic time, and temperature changes that give the earth ice ages. These events recharge the surface with elements necessary for life, and once those elements are present, they can combine with living things to make soil. Europe and North America had mountain uplifts and the repeated glaciations of the Pleistocene. Uplifts recharge the minerals. The grinding ice of the Pleistocene glaciers pulverized the rock, releasing those essential elements and setting them loose in the biota, where nosing roots would capture them to combine with their atmospheric relatives—carbon, hydrogen, oxygen, and nitrogen. We are major beneficiaries of this ice, which came and went over this nearly two-million-year period.

Whereas we have been fortunate, Australia has not. Its last geologic activity was sixty-five million years ago, and that continent is likely to stay relatively poor for a very long, probably beyond human time. With such a poor nutrient base, its standing crop of life will never weigh as much as that of the United States, even with comparable precipitation—not until a geologic recharge event. From this perspective, soil is as much of a nonrenewable resource as oil.

An Appeal to the Russians

Near the end of August 2009 in Coon Rapids, Iowa, there was a cel-
ebration of the fiftieth anniversary of Nikita Khrushchev's visit to
the farm of Roswell Garst, who raised and sold hybrid seed as well
as farmed. Sergei Khrushchev, Nikita's son, now a US citizen, was
there, as he had been fifty years earlier with his father. There were
twenty-five or so Russians present, including the Russian minister of
agriculture and the ambassador, plus a couple of hundred Americans,
including representatives of US agribusinesses. I was asked to give a
talk. I accepted, thinking this a good opportunity to invite the Russians to a cooperative effort on the substance of the 50-Year Farm Bill,
a proposal I had launched with Wendell Berry and Fred Kirschen-
mann to promote gradual, systemic change in US agriculture.[3]

In preparation, I had read some of Khrushchev's memoirs, noting
especially the time of the visit with the Garst family plus much that
had happened since. I remember that visit in 1959, but with the mind
of the twenty-three-year-old graduate student I was then, aware only
of a few bare bones of history. I remember that Khrushchev met with
President Eisenhower, that he crossed the continent, had a dust-up or
two, and visited farmer Garst in his Iowa cornfield.

Another important historical marker: 1959 was the fiftieth anniversary of what energy scholar Vaclav Smil has called the most important invention of the twentieth century, the Haber-Bosch process, a process that relies largely on natural gas as the feedstock to
turn atmospheric nitrogen into ammonia fertilizer. The yields in
corn and other grains began to increase greatly around 1959, largely
due to the application of that industrially created nitrogen fertilizer.
This increasingly industrialized agriculture was the background for

---

Excerpted and adapted from Jackson, *Consulting the Genius of the Place.*

Khrushchev's visit to Iowa. What were he and Garst thinking as they stood and talked? What was the entire delegation thinking?

The one group knew starvation, some of them maybe firsthand. Only fifteen years earlier, the siege of Leningrad and the resultant starvation, sickness, and death had lasted from September 1941 until January 1944. No matter how long and dark the tunnel might be for the countless problems humanity faces, both farmer Garst and politician Khrushchev knew that if we could produce enough food, most other problems would be manageable. Their question then is our question today: How do we help ensure an adequate food supply, not just in our two countries but around the world and for centuries to come? When Garst and Khrushchev met, the world population was approaching three billion people; we are now coming up on eight billion.

Were Khrushchev and Garst, standing together in that tall cornfield, aware that those US soils had been made rich by the glaciers of the Pleistocene? I imagine that neither the capitalist nor the communist brought up the geographic differences between the two countries. I doubt that they pondered a 1.8-million-year-old Pleistocene product—no, a gift—where those North American glaciers had scraped the soil off the Canadian Shield and deposited it right under where they were standing. The same gift came in places such as Ohio, Indiana, Illinois, Wisconsin, Michigan, and Minnesota, even northeast Kansas, northern Missouri, and parts of Nebraska and the Dakotas. That is the core of the source. And there was the somewhat predictable moisture coming from the Gulf of Mexico to irrigate corn, the most productive grain of all.

Both were likely thinking about, and both may have been arguing for, the virtues of their respective political systems, ignoring the geographical origins and the fact that something like two-thirds of Russia's latitude lies north of the northern tip of Maine's latitude. But of the six relevant continents, none have the agricultural potential of North America. The Cold War was underway and growing. Both empires wanted to show their stuff, and that included agricultural production.

There is another matter. Neither Nikita Khrushchev nor Roswell Garst nor most anyone else could imagine the speed of change to come over the next fifty years. How could anyone have imagined the scale of the consequences of burning the manure of the farmstead and substituting chemical fertilizer, as Mr. Garst advocated? Mr. Khrushchev thought the idea a good one. Few of us, I'm betting, could have foreseen the number of nitrogen-polluted wells from commercial fertilizer in rural America or the cost for cities such as Des Moines of removing nitrogen from the public water supply. How many could have foreseen that nitrogen runoff would cause numerous dead zones in the oceans?

By most accounts, the Industrial Revolution was a bit over two centuries old when the still-early stages of agriculture's so-called Green Revolution were getting underway in 1959. But that was about to change fast. Within three decades, yields of several major crops had doubled in some places, tripled in others. Oil supplied the fuel for traction and other mechanical operations. Natural gas sponsored the ammonia to fertilize the crops that realized the yield potential developed by the plant breeders. Every gain in bushels per acre or kilograms per hectare was praised. As the food supply went up, antibiotics and other drugs forced death rates down. By 1999, the world population had doubled. By means of all these successes, what we can call the Industrial Mind now permeates the world. As fossil-fuel consumption accelerated, only a handful of scientists speculated on the costs to the ecosphere of the released carbon. But now global warming is generally recognized as a public enemy.

Nikita Khrushchev and Roswell Garst could not have foreseen the future we now live in, but they did know one big thing, as true today as it was then: People will want to be fed every day, every week, every month, every year. This need must have been most poignant for Premier Khrushchev.

Solutions to soil erosion, as well as the energy and climate crisis, will require extraordinary political will, both to conserve topsoil and to stop climate change and develop renewable energy technology. We might succeed on the energy front, but once soil has eroded, its restoration comes in geologic time, and no technological substitute will do. And in spite of all our efforts so far, soil erosion and other landscape degradations are increasing globally. In a few places, they have been slowed by minimal-till or no-till farming, but with this so-called fix, pesticides accumulate. We are poisoning our soils to save them.

With all of that in mind, in my talk at Coon Rapids, I proposed a solution that both of our countries could embrace, believing that if we cooperated, success would come faster. Here are the highlights of what I said that day.

I described The Land Institute's mission, explaining that while annual grains grown in monocultures dominate agriculture, essentially all of nature's ecosystems feature mixtures of perennials. We need to breed perennial grains, which would amount to "new hardware," allowing us to draw more heavily on the literature of ecology and evolutionary biology. This literature has been accumulating on the shelf for a century and a half, but for its own sake. That knowledge represents great untapped potential. Ecologists have learned to detect many of the efficiencies of various ecosystems, both above and in the soil. The ecosystem becomes our conceptual tool. Nature's way of operating on the land becomes the standard against which we judge our agriculture.

Imagine two ends of a spectrum—human cleverness at one end and nature's wisdom at the other. The Industrial Mind, over the past one hundred years, has increasingly relied on human cleverness. I am not proposing that we quit being clever; instead, cleverness should be subordinated to nature.

Implanting this idea in our minds will be a great challenge. Our ten-thousand-year history of growing food has been tied to the notion that nature is to be subdued or ignored. But this attitude has led to relentless deficit spending of the earth's ecological capital. Now, finally, we face the need to change course. The possibility of doing so resides in exploring the efficient processes of nature, which are sponsored by contemporary sunlight. This look to nature begins with the soil itself. For tens of millions of years, nature's arrangements have managed the twenty-some elements that go into all organisms. Only four of these elements—carbon, hydrogen, oxygen, and nitrogen—are found in the atmospheric commons.

I explained why Russians need to be major players in this effort for a new agriculture. I told five stories from Russia's rich scientific history. Three were about scientists, one about a moment in history, and the fifth about a historical period. I used all five stories to lay the groundwork to explain what I hoped our two countries could accomplish over the next half century.

Here is the first one: In the 1870s, Vasily Dokuchaev, the father of soil science, was given the task of describing the structure, origin and evolution of the deep, rich grassland soils of western Russia. Classifying them had been elusive. The dominant belief of the time was that weathering alone was responsible for soil formation. Soils were thought to have no emergent properties of their own, no properties due to interactions with organisms, nothing that would give them standing beyond what mere weathering would cause.

To accomplish this task of soil classification, Dokuchaev traveled over ten thousand kilometers, about six thousand miles. From his observations, he concluded that "soil exists as an independent body and has its own special origin and properties unique to it alone." More is involved in soil formation than merely moisture and temperature. He identified five factors that govern soil formation: climate, parent material, organisms, topography, and time. With his conclusions, soil science was revolutionized, making him in some way the founder of the discipline. Lacking translation from Russian, Dokuchaev's ideas were not recognized in the West for decades. But when Professor Hans Jenny, a central figure in American soil science, published his seminal book, *Factors of Soil Formation*,[4] in 1941, he accepted and used Dokuchaev's five factors. With the use of differential equations, Jenny created a formula, providing quantitative measures of the five factors of Dokuchaev and kicking off a revolution in soil science. That is number one, and that's big. Now for the second big name.

One of my most-treasured photographs on a shelf in my office is of Nikolai Ivanovich Vavilov, who was born in 1887. It was taken in the 1930s and was given to me by botanist Gary Nabhan. In the photo, taken on the Papago Reservation in Arizona, Vavilov is looking at the herb devil's claw, which he recognized as a cultivated plant. This great scientist was internationally known as an agronomist, botanist, plant breeder, geneticist, and plant geographer. He had a big-picture view of our ecosphere, both geographically and in time.

Early in his career, he set out to determine the centers where cultivated plants originated, traveling worldwide and making massive collections. His published conclusions set the standard for all subsequent investigations. Plant breeders still turn repeatedly to these centers, though Dr. Jack Harlan updated some of Vavilov's conclusions. His passion and intellect were supported by a rich Russian culture notable for its love and honoring of natural history. There is a very dark history of how he was treated by his country, but that is another story.

The third great man is Theodosius Dobzhansky, born in Ukraine in 1900. He immigrated to the United States, became a citizen, and is regarded as the most important evolutionary biologist of the twentieth century. In 1937, twenty-two years before Khrushchev's 1959 visit, Dobzhansky published his landmark book, *Genetics and the Origin of Species*,[5] in which he bridged a wide gap between experimental geneticists and naturalists. He was the first to successfully integrate the understanding of evolutionary problems from his naturalist perspective with experimental genetics. The late American evolutionary biologist,

Nikolai Ivanovich Vavilov with the herb devil's claw in the 1930s. Photo gifted to the author by Gary Nabhan.

historian, and philosopher Ernst Mayr said that Dobzhansky's 1937 synthesis, so long in arriving, was the most decisive event in biology since 1859, the year of Darwin's *On the Origin of Species*.

Now for the other two parts of the Russian story I thought important.

During the 872-day siege of Leningrad, hundreds of thousands of people died from hunger. There was every reason to believe—given its huge quantity of seeds collected from all over the world—that the Institute of Plant Industry would be overrun by hungry people. But somehow or other, the institute's scientists and technicians guarded the collection. With seeds all around them, they, too, starved. The collection remained untouched. I find that a moving tribute to a country's tradition of commitment to science.

Finally, in the past extensive work on the remote hybridization of plants has been carried out in Russia. A team of plant breeders made countless wide crosses between species and varieties out of a desire to speed crop plant evolution. According to N. V. Tsitsin, whose name is best known, one of their efforts was to develop a wheat with "perennial character, remarkable for dwarfness, resistance to lodging, an ear structurally similar to that of conventional wheats, and easy threshability." He also said that "among the primary features of the perennial and feedcorn wheats we are currently putting through selection is *their ability to develop a powerful rooting system* (emphasis added), a factor rather important for the maintenance and betterment of the soil structure, but one of prime importance for the regions susceptible to wind erosion."

Tsitsin acknowledged that "it will be some time before the newly bred varieties can be cleared for production, lagging far behind the best wheat varieties in terms of their yields." He went on, "The fields under perennial wheat need not be reploughed, the stubble will be highly snow-retentive and thus very likely to promote the accumulation of soil moisture. In sum, these factors will provide for *sustained and progressively increasing fertility of the fields*" (emphasis added).[6]

Of course, one wonders what happened with the Russian effort. Why did the work on perennial wheat not continue? It was merely part of a larger program in the Soviet Union, pursued more or less as a sideline, because (1) farmers and governments need ensured high yields every year, so the bulk of research funding went to that end, and (2) any long-term research effort in agriculture is first to be cut when funding is reduced.

Achieving high yield in perennial grains requires long-term effort. For the moment, fossil fuels make it possible to mine, package, and transport nutrients from afar. And, as noted earlier, commercial nitrogen fertilizer is made possible by the energy-intensive Haber-

Bosch process. For whatever reason, the perennial wheat breeding faded away before it was farmer-ready in the Soviet Union.

From Russian and Soviet history, back to the challenge today. In the 2009 meeting, I talked about the practical necessity for a joint venture with the Russians, hoping those five stories could bring us to face the modern challenges of achieving sustainable food production—to feed the expanding population as we work to stop its growth, prevent soil erosion beyond natural replacement levels, manage nutrients and water more efficiently, and greatly reduce the use of toxic chemicals. I did not mention the need to use perennial polycultures as a way to reduce greenhouse emissions but should have.

I thought that what I said was enough motivation for our two countries to join hands and undertake a massive breeding program devoted to perennializing the major crops currently responsible for occupying at least two-thirds of the agricultural land of the planet and responsible for about two-thirds of our calories—mostly grains and pulses (legumes). Joint ecological studies of agricultural landscapes could get underway at the same time. Just in case they were suspicious, I emphasized that to this common effort, The Land Institute would offer free germplasm and more than thirty years of experience with perennials. Our staff in that decade had already greatly enhanced the diversity of crops and increased the speed of change. We had hybrid prototypes we could draw upon.

After the polite applause for my presentation, Sergei Khrushchev, now an American citizen on the faculty at Brown University, spoke, and then all two hundred of us had lunch on the Garst lawn with the same menu as fifty years earlier. The food was great. It was a beautiful day.

At the end of lunch, Sergei Khrushchev pulled the Russian minister of agriculture and me aside as the crowd began to thin. I extended my plea for a cooperative effort between the US Department of Agriculture and the Russians because of their rich history. Through the translator came a response for which I was not prepared. The minister of Russian agriculture said, "I need to be blunt with you. We don't have the scientists." I said, "Well, yes, I know about the great purge in genetics during the Stalin era when some three thousand were forced out of genetics work because of Lysenko."

"No, no," Sergei Khrushchev said. "Not Lysenko, Yeltsin."

He was talking about the Russian economic collapse and brain drain in the decade after the collapse of the Soviet Union. We exchanged pleasantries, exchanged cards, and shook hands, and the scientists all left, Sergei with them.

By two in the afternoon, my wife, Joan, and I were driving home, both ruminating on what had transpired, beginning with the reception and dinner the night before. It would have been easy to focus on the consequences of the failed Soviet system and US superiority. But the world is complicated, and I thought back to the era in which Mikhail Gorbachev presided over the end of the Soviet Union, followed by Boris Yeltsin crashing through the 1990s. I remembered an Associated Press article I had read in 1997 about a Russian welder in Siberia.[7] At home, I reread that article:

> Each day, Nikolai and Galya arise in the dark and go about the business of making a living. They milk their cows, feed their pig, gather eggs from their chickens, tend their garden. They live off what they grow, and sell the rest for a few rubles here and there. From milk alone, they earn perhaps $100 a month. And when the sun rises, Nikolai heads off from his simple wooden house to his long-time job as a welder in a state-run auto repair factory. For this, he earns nothing.

The article continues, "People survive on their gardens and their wits, and the official economy primarily is a distraction." After some mention of an impending trade-union strike and President Yeltsin's concern about doing something about it, the writer says more:

Across Russia, especially in smaller towns and villages, millions of workers have gone months without wages. Both the government and private employers have been unable—or unwilling—to pay them. Even retirees have gone without their pensions. Outsiders tend to ask how this is possible: How can a nation survive when its people are unpaid? Why would a worker show up for a job that offers no wages? Like many Russians, Nikolai—who hasn't been paid in three months—doesn't ask these questions. Why wouldn't he show up for work? "Where would I go?" he said. "There aren't any other jobs in this town. I'm too old to look for work in Moscow. This is a one-factory town; we have no other choices. And besides, what if the day I decide not to show up the managers start handing out wages?"

Though not explicitly mentioned in the article, it was not the industrial economy but nature's economy, in combination with traditional culture, that had made this family resilient and continued to feed the people, even subsidizing the stumbling industrial economy.

Imagine nearly any average US citizen, except for the Amish and a few other outliers, going with no wages in the United States for three months now that our traditional rural economies have been mostly undone. (As I write this, the COVID-19 pandemic is creating exactly these conditions, with dire results for many in the short term and uncertain long-term consequences.) The collapse of the Soviet empire represents the first major failure of the Industrial Mind. But both of the competing Cold War systems concentrated power and in so doing greatly reduced the number of people on the land and in small communities.

What is needed for all of us here—for us, the Russians, and the rest of the world—is to increase our imagination about what real resilience means while we still have slack.

The Lilac Tree Is in Near-Full Bloom This Morning, but So What?

Anytime I see anything written by my late friend Harold J. Morowitz, I want to read it. Sometimes I find myself circling back to reread. Harold was a professor of molecular physics and biochemistry at Yale and later at George Mason University. He died in 2016 on his way to teach a class at the age of eighty-eight. I recently came across a quotation I had marked long ago in Harold's book *Cosmic Joy and Local Pain*.[8] He was quoting Steven Weinberg's book, *The First Three Minutes*. Here are Weinberg's words:

> It is almost irresistible for humans to believe that we have some special relation to the universe, that human life is not just a more-or-less farcical outcome of a chain of accidents reaching back to the first three minutes, but that we were somehow built in from the beginning. As I write this I happen to be in an airplane at 30,000 feet, flying over Wyoming en route home from San Francisco to Boston. Below, the earth looks very soft and comfortable—fluffy clouds here and there, snow turning pink as the sun sets, roads stretching straight across the country from one town to another. It is very hard to realize that this all is just a tiny part of an overwhelmingly hostile universe. It is even harder to realize that this present universe has evolved from an unspeakably unfamiliar early condition, and faces a future extinction of endless cold or intolerable heat. The more the universe seems comprehensible, the more it is also pointless.[9]

Though it was several years before that I had marked that passage, I was just as bewildered on my latest reading as when I had first read it. Harold pointed out that it is ironic that the very scientists who consider existence pointless "have spent lives working hard to make a point, and their very lives are examples of meaning in our finite world of human culture." Is there not some paradox here?

It is not surprising, given his Nobel Prize and ability to explain science to a lay audience, that Weinberg's writings receive a lot of attention. As I was wondering what I was supposed to do with that claim that we are a "farcical outcome," my wife, Joan, called me to look at and admire our lilac tree, which was in near-full bloom that morning. We'd had lots of rain that spring, and as a consequence, we were being treated to the most spectacular display of leaves and blooms in my memory of this place. Our arboretum trees were larger and blooming, the grasses more luxuriant. Diversity is on the increase. For example, broad-leafed flowing plants called forbs are accumulating in the small prairie patch and in the cow pasture as well. Are we supposed to assume that all of this wonder, some planted by us but most produced by nature, is—to use the last word of Weinberg—"pointless"? Yes, from one perspective, we do live in an "overwhelmingly hostile universe." Does that make it pointless?

In his book, Harold wrote about several scientists, including Weinberg, who gave us so many insights into our origins during that last three hundred years of our journey. What about Isaac Newton, who elucidated the laws of motion and theories of universal gravitation the help us understand how bodies interact? As Morowitz said, because of his insights, we now understand those complex motions in our own solar system, using some of the fancy math he invented. What is pointless or meaningless about that?

Harold continued with the listing of scientists after Newton, when others came along and gave us formal insights into thermodynamics, statistical mechanics, electromagnetism, laws on radiation. Is all of that without meaning?

Harold referred to the "enormously powerful concept of temperature," without which it was not possible "to develop the earth sciences in terms of their physical foundations." I was so struck with that comment that I awoke in the night and sat down with my iPad to look up the history of the development of metrics for temperature. Several were involved beyond the familiar Fahrenheit and Celsius. It makes one wonder where we would be without the simple instruments that allow us to get accurate readings on heat.

Staring at the ceiling and thinking of other ways humans have estimated heat, I remembered my college summers as a welder and shaper of metal and thought of certain modern farmers who repair their own equipment using a torch, vice, and clamps, still relying on color, as we did on the industrial floor. Seeing when the heat is right made us and the Iron Age people one, for they had to rely on the color of the ore and metal to practice their art in the making of instruments for domestic life and, of course, war. Such work without the modern thermometer was not pointless, and neither is the thermostat with its bimetallic strip that turns the furnace off and on to regulate heat for the home. None of that is without point or meaning to me, or, I imagine, to most people who think about it. Are we not thinking deeply enough for Weinberg's way of thinking? Is it too mundane?

Harold Morowitz cites Einstein's theory of relativity and all those early insights in the last century that fit together in such a way that physicists came up with models, not just of our galaxy but also of our universe on a scale that is still incomprehensible. Because of Marie and Pierre Curie and Henri Becquerel, our knowledge of radioactivity and nuclear reactions expanded to the point where our own kind began to explore how stars are born. How is it that they can shine for billions of years, but their destiny is still to grow old and die? Thanks to quantum mechanics, we can now understand all that better, as well as how distant stars are from us and from each other, based on the spectrum of light.

New nouns have been required in this journey. I was reminded of that in reading the *New York Times* obituary of Murray Gell-Mann, who predicted the existence of quarks and other subatomic particles. He was not yet ninety when he died in May 2019, which reminded me of how recently this knowledge has emerged. How is any of that meaningless? It is all something to know, whether it is practical or not.

It is mostly here at home on our creative Earth, where our interests seem to soar, beginning with the journey on early Earth. We have a framework now to help understand the journey from minerals to cells. Indeed, we have more than a mere framework on Darwinian evolution and epigenetics. All of us carry within something shared with the oldest of all aerobic creatures: the citric acid cycle (the sequence of chemical reactions through which aerobic organisms generate energy). Morowitz calls this the oldest living fossil. How wonderful to know that as we stare at a small ant. Does it not make us one with it, the whale, the milk cow, a California redwood, and all the rest of aerobic life?

We all have to confess—no, not confess but simply acknowledge— that there will probably never be a point at which we mere humans will have it all nailed down. But so what? Does that make it pointless? How much more do we need? Is it not enough to simply say that we are matter and energy's way of having gained self-recognition? What if we do discover whatever was behind the Big Bang? What might it be, more universes? Are we not still left with the question, the big one: Why anything, why not nothing?

Testable predictions have been made and likely will continue into the foreseeable future. It is close to being common knowledge that our universe is some 14 billion years old, our planet about 4.5 billion years old, and we *Homo sapiens* with the big brain around 200,000 years old. Wonderful to know, eh? When Steven Weinberg speaks of the "overwhelmingly hostile universe," so what? That ain't now, not here. And when we face "a future extinction of endless cold or intolerable heat," so what again?

And when he adds—here is that last sentence again—"The more the universe seems comprehensible, the more it is also pointless," I wonder, what point does he want? The very fact of our existence seems enough. That I am called from my nonelectric typewriter by Joan to come and look is not pointless. It is good enough for me, for my arrangement of former stardust, at least on that day, to have had the pleasure of talking to my ninety-one-year-old brother, the last of my five siblings still living. He interrupted my typing to see how I'm "a-doin'" and to talk about matters of family interest. Pointless?

Weinberg clearly doesn't act as if he thinks his life and work are pointless. And so, to be fair, it's possible that by those comments he meant something like this: "The more we learn about how the universe works, the less we should need supernatural explanations attributed to a divine entity, and one consequence of that is that we have to let go of the idea that meaning is inherent in the universe. What we are left with is the meaning that we create." If that is what he meant, I won't argue. But I don't call such a world pointless.

Dear reader, you can take it from here. I'm going to go look at that lilac tree again. Maybe it has opened up even more and is even more spectacular. And on the way, I just might make a loop to admire the yuccas in bloom, and the blue wild indigo, and the eastern gama grass, and the always-ready-to-pounce calico cat, Susie, and the flow of water on the river over the Wellington shale, and an "about-right temperature" on this thirty-acre slab of space-time for which I have the deed, wondering how I can claim ownership when I am the one who is owned. Gotta go!

Ideas I've Run into along the Way

It can be dangerous to think too much.
But it's even more dangerous to think too little.

The Danger of Nuance without News

It is an old theme here at The Land Institute, to be mindful of the necessity to have the boundaries of consideration overlap the boundaries of causation. These terms, which I spotted years ago in *The Dialectical Biologist* by Richard Levins and Richard Lewontin,[1] remind us that if we want to know how things really work (the question of causation) we have to think as expansively as possible about how a whole system affects outcomes (the scope of what we take into consideration). The idea helps us be mindful of the various slabs of space-time (an important term for me, borrowed from ecologist J. Stan Rowe) that we want to discuss or explore.

At The Land, we also are mindful of history and the need for a "historical imagination," a term I first encountered in a 2012 *New Yorker* article by Adam Gopnik.[2] Lots of people show some interest in history, but it's the job of professional historians to walk the line between "scholarly rigor" and the need to keep the "historical imagination" alive. Gopnik wanted his readers to think about the risk of turning history into "nuance without news." He wrote, "The pursuit of scholarly rigor too easily leads historians to erase any signs of the historical imagination from their work." He described the historical imagination as the "ability to see small and think big." Acknowledging that while only thinking big can lead to "melodrama and fantasy," the opposite is just as bad—only seeing small "makes you miss history altogether while seeming to study it."

For emphasis, Gopnik argued that "any significant change in human consciousness can be dissolved if you break it down into its individual parts, which are bound to seem contradictory or many-sided." For example, the concept of an Italian Renaissance is a way to explain a period of time—that's the thinking big—built by many acts of seeing small, the patient work of description of specific people, institutions, systems, events.

My friend the historian Angus Wright had another way to talk about this need to see small and think big. One year at our annual Prairie Festival, in response to a student's question about how to approach university studies, Angus suggested that intellectual efforts benefit from a "focus in/focus out" approach. One can start by looking at a very specific question. The focus moves in close. But if you stay at that level forever, you don't know how to see the patterns in what you learn, Angus said. That's the time for the focus to move out to search for a bigger picture. Throughout one's intellectual life, Angus said, it's good to focus in, focus out—toggle between the two.

We've all known people who are satisfied by learning lots of details and telling others everything that they've learned (Lord help you if one of those folks corners you at a party). We've all known people who can't be bothered with details because they have a grand story to tell that connects everything. Angus is right—don't get stuck at one level.

Wondering about the Origin of Fire

The title has little to do with the point of the story, but it helps get us started. One January evening, I was watching the fire burn on the other side of the glass in one of our woodburning stoves. I had been fooling with the stove damper to get the right amount of air, enough oxygen to get a good fire, and then, once it was going well, to adjust it to an acceptable level. Back in my chair and admiring the flame, I wondered when was the first open-flame fire on Planet Earth? That was the first time I asked that question. And then I wondered, how is it that it me took so long to ask it? I am in my eighty-fourth year, for gosh sakes. Didn't we all learn as kids that for something to burn, it requires oxygen, and that we need oxygen to breathe? Sometime, probably after high school, I learned that the early earth had next to no free oxygen. And so the question: When did enough oxygen accumulate to support a fire with an honest-to-god open flame? Joan and I looked it up and found that the oldest evidence goes back to 420 million years, when a small, leafless plant from the Silurian geologic period showed char remains in English siltstone. So we have a date.

But still the question: Why did it take me so long to wonder when the first flame arose? There were several disciplines I could have called on, so it wasn't the lack of knowledge. It must have been something else. What that "something else" is remains a mystery. I will have to leave it at that, why we take some things for granted and don't think to ask.

The origin-of-fire question got me to thinking again about the different fields of science that have worked together successfully to learn things. How much scientific knowledge about the history of the earth was necessary to have a satisfactory answer to my initial question? Of course, it is sure to vary from individual to individual. But here is a start. Without photosynthesis, we would not have fire. Photosynthesis is the result of light coming from the sun. We all know that. Given the speed of light, we believe it takes about eight minutes and twenty seconds to make the trip. (Notice I did not say, "We know.") We do know that once sunlight strikes the leaves of plants that contain chlorophyll, energy capture in the form of photons happens.

Nearly every schoolkid, some time or another, is exposed to that fact, but it took a while to know enough of the whole story of how the process works. Without knowing the whole story, the chemistry that happens is seemingly magical. Here is the elementary sequence: The chlorophyll has the ability to take carbon dioxide and water and turn it into a sugar that includes newly arrived energy from the sun's surface. In the process, the plant exhales oxygen, in fact two oxygen atoms we call O_2. Now, for a "balance of nature" reality: Atmospheric oxygen can return to a sugar molecule and release the energy to do work. That process is called respiration.

I have told the reader nothing new to most mildly educated people. So why tell it again? For this reason: Something was missing from our understanding of the entire chlorophyll story. It was pretty late in the detailed understanding before scientists figured out how the light coming from the sun was converted into stored energy. And here is the point: Biologists who studied such matters did not have an answer. I am among those who have forever been talking about the problem of reductionism in science, which dates back to near the beginning of science and the Enlightenment. It is also fair to say something like "Well, yes, that is a problem, but at the same time, it is worth being mindful that disciplines do frequently come together."

Some time ago, I read a review[3] of a book by Peter Watson, *Convergence*.[4] The reviewer, Joseph Swift, acknowledged how multidisciplinary approaches often require a common ground and that finding a "shared territory" has not always been easy. He gave an example from the late nineteenth century, when some biologists wondered whether a "vital force" was necessary for life. That would make the creationists happy if it were so. They actually investigated this. While they were at it, some physicists were not far from discovering the electron. Swift wrote, "At that time, it would have been an enormous leap to suggest that energy from electrons could be harvested by plants, and that this energy was part of the mysterious force biologists were searching for."

So here is a beautiful story for the history of science. Filling in the blanks required quantum physics scientists. But, and here is the big but, their physics knowledge could not do it alone. The chlorophyll molecule resides inside chloroplasts, which, in turn, reside within the cell. Cell biology is a field. Another field called biochemistry was called upon. Something closer to a total understanding of how light is converted into stored energy, well, that required the quantum physics scientists—with knowledge outside the boundaries of the two other fields.

Several disparate sciences have increasingly come together, raising new questions.

Twelve years after Watson and Crick elucidated the spiral-staircase structure of DNA, I took an exciting biochemical genetics course. The course was at once interesting and helpful for understanding how the genetic code worked all the way back to the biochemical physics levels. I am not keen on all of the necessary disciplines, but it was enough to integrate ecology, genetics, biochemistry, and physics (four disciplines) and develop a framework, even without all the details.

Now, back to the fine fire behind the glass of our woodburning stove. Does the question about "when was the first fire" have any value? Does knowing something that is only wonderfully interesting have value? Is this a necessary something if we are to reverse the drawdown of the long-term ability to support a variety of life and culture? If the first piece of that solution is affection for such knowledge—and all of us graduate students seemed to have it in that molecular genetics course—maybe, just maybe, the experience begins with wondering and then finding out something wonderfully interesting about Earth, our mother and defender, like when the first fire was lit on what became our home. What do you think? Is it not our job to see to it that "soul enrichment" can also have utility?

How Lothar Convinced Me that Dinosaurs Did Not Exist
until Humans Discovered Them

LEXINGTON, Ga. (AP)—An elderly woman killed by a pack of wild dogs had been out for a walk when she was attacked, and her husband died trying to fight off the mauling animals when he discovered the bloody scene near their rural Georgia home, authorities said Tuesday.

Preliminary autopsy results showed Lothar Karl Schweder, 77, and his 65-year-old wife, Sherry, died from multiple animal bites.

That August 2009 news report hit me hard. I knew Lothar as both a friend and a colleague. I had first met him more than four decades before.

Born in Poland in 1932, Lothar came to the United States after World War II and eventually became a professor of German and philosophy at the University of Georgia. His interests were broad, particularly in history, philosophy, and science. He was a fascinating conversationalist. I knew his wife, Sherry, only slightly; she had been a humanities bibliographer for the library at the University of Georgia.

Learning of his death reminded me of an extended, and to me very useful, conversation I had had with Lothar. It had to do with our perception of reality and its sources and how it can change with history, time, circumstance. I knew all that, I thought, but I was having trouble with one of his contentions. Somewhere during the argument, I thought he had pushed his argument to a level of absurdity, and in frustration I pulled out the stops and said, "Well, hell, Lothar, dinosaurs were here before people."

"Well, Wes," he said, "as a matter of fact, dinosaurs were *not* here before people. That is what I have been trying to tell you. Dinosaurs *as such* were not here until people began to discover them."

It was the "*as such*" part of that sentence that made me begin to understand a bit of what he was having trouble making me understand. It was about the nature of perception and how our world changes with our discoveries and therefore our changing perceptions. The individual animals we call dinosaurs lived and died before us, of course, but the category and concept of a dinosaur are human creations.

We can all acknowledge that revolutions in thought have happened and still happen. We talk about worldviews and how they change. They may come from a new body of data or from someone's new interpretation. Sure enough, in the last score of years or so, we have acquired a new understanding of the universe. Orbiting telescopes are now available for astrophysicists and astronomers to take a more detailed look, and when they explain it, the story becomes much different. The universe is much larger than previously believed. The dynamic interactions are different. Taken all together, before satellite astronomy emerged, the cosmos *as such* in our minds did not exist.

Now, those who have spent some time in certain academic circles might be asking, "Has Wes gone postmodern?" Rest assured, I have not, at least as far as I can understand the term. For those who are outside those certain circles (and you probably should be grateful for being outside them), the term "postmodern" may be unfamiliar. Here's what usually trustworthy Wikipedia has to say: Postmodernism is "an attitude of skepticism, irony, or rejection of the grand narratives and ideologies of modernism, often calling into question various assumptions of Enlightenment rationality." The most extreme form of postmodernism, caricaturing it only slightly, would be the belief that for us humans, there is no reality outside our own discourse. Of course, everyone knows there is a material reality not dependent on humans—the water of the Pacific Ocean will be there no matter what we have to say about it. But rather than poke fun at theoretical debates in the humanities, I want to take seriously what Lothar was trying to get me to understand.

As someone trained in the sciences, I have no problem with "Enlightenment rationality." But as an ecologist and critic of industrial agriculture, I have long discussed the problems caused by science's reductionism and the modern world's embrace of high energy/high technology. Enlightenment rationality has led us to believe we know more than we know and can control what we cannot. We should be wary of "grand narratives" that purport to tell the whole story, although I remain persuaded of the argument for Darwinian evolution by natural selection, one of the grandest narratives we have.

But back to Lothar. While I see his point, the hardheaded realist in me still wants to ask, "What does that have to do with the predicted yields for the Kansas wheat crop this year?" Or, in the following example, the sugar-beet harvest. This comes from a "Food and Values" conference I once attended, where I learned about food being used as a weapon. Our government subsidized sugar beets and stood ready to dump sugar on the global market when Fidel Castro—the communist whom so many US leaders loved to hate—got out of line in Cuba. So now, when the topic of sugar beets comes up, I think of them as a crop grown by beet farmers but weaponized by our government. Nothing changed except my perception.

That's a simple example of how learning new things changes our view of the world. But sometimes the effects of our ways of seeing the world are not immediately evident and are not a matter merely of individual understanding. We might say that there is a witless inner penetration in countless decisions. Our values affect the genotypes of our crops and livestock. There are Chicago Board of Trade genes, computer genes, fossil-fuel wellhead genes in our crops. None of those exact DNA arrangements would exist if there were no Chicago Board of Trade, advanced technology, or oil and gas.

Once we start pondering a large suite of questions, all the world changes. When I see a grain truck at the elevator or being filled in the field, do I see a weapon? Yes, sometimes. When I see a near-weedless field of corn or soybeans, I know it likely has been sprayed with the chemical herbicide featuring glyphosate, the key ingredient in Monsanto's Roundup weed killer. My mind turns immediately to the growing number of researchers who have concluded that that chemical is a probable cause of non-Hodgkin's lymphoma. Such near-weedless crops no longer attract my admiration. Instead, my imagination goes to the soil of those fields, the organisms within, and the groundwater that supplies farmers' households, cities, and towns. I think of farm families who "walked the beans" with a hoe. Those fields are no longer what they once were—*as such.*

Back to Lothar and Sherry. According to newspaper reports, the dogs that killed them were all known by the neighbors. None of the dogs had shown any sign of aggression. According to the county coroner, "The animals did not seem overtly threatening but were guarding the bodies as if they were prey."

At the moment of the attack, at least one, if not all, of the dogs *as such* was a killer. Standing over the two human bodies and before, they were at that moment not pets, even though they had been domesticated companion animals just minutes before. The animal-control officers said the dogs were neither malnourished nor rabid. Officials had never received any complaints about the dogs.

It is no secret that both people and dogs evolved as predators. Obvious testimony to this is that our eyes are at the front of our heads. Were we vegetarians, as are cows, our eyes would be at the sides to alert us to any predators approaching from behind or from the side.

What is this all about, anyway? Of course dinosaurs were here before humans discovered them. "OK, I get it," you may say. "Dinosaurs were not here *as such* until humans were here. I get that, too." What are we supposed to do with the distinction when both are true? Is this an invitation to some cultural revolution?

Well, yes, it is, knowing that revolutions in culture and science have come and will continue. Some ways of thinking have gone away. I don't think the basics of Darwinian evolution will change much. Genetics has been somewhat modified with epigenetics (the study of how genes are expressed in an organism, which can change depending on the environment). Will creaturely evolution have to be rethought as we ponder the role of viruses and marine microbes in evolution? Even without dramatic changes in our knowledge, some very basic questions persist, and changes in how we answer them might end up promoting dramatic shifts in perception. Evolutionary biologists have been asking for some time, what is a species? For that matter, what is an organism? The problem of classification is old, and scientists come back to it again and again. Who knows when something big might break?

So am I making too big a deal out of this? In our daily lives, we know that it is not unusual for people to see things differently—sometimes dramatically differently. Does it matter what anyone believes to be reality? Sometimes it does, sometimes it doesn't.

When the various Caesars sent their ships from one end of the Mediterranean to the other, it was, de facto, a Roman lake. For the Romans, at one time there was nothing beyond the Straits of Gibraltar *as such*. Those rocks at Gibraltar were called the Pillars of Hercules. Written in stone were the words "Ne Plus Ultra!," which means "no more beyond." This Roman lake made it easy to get around and travel to one cultural shore after another. The Romans' horizons were broadened and in turn helped define much of the Roman world, though still limited by the Mediterranean *as such*. The Pacific Ocean, *as such,* did not exist for them. North and South America did not exist *as such*. For that matter, North and South America *as such*—as continents—did not exist for the people who lived there at that moment, who had not sailed to other land masses. We are all products of that period, which is part of the history of the Renaissance, the Reformation, the Spanish Inquisition, the Enlightenment, all of this as a forerunner to the Industrial Revolution.

Lothar's point is well taken. Lives change when perceptions change. And now you may be saying to yourself, "For gosh sakes, Wes, don't we know all that?" My answer is, "Yes, we do." We can under-stand how a household pet becomes a predator for the first time. What about this: Lothar, a thoroughly decent person, had been a member of the Hitler Youth. Did he have a choice in that? How can we best answer that without further questions and more answers? What did membership do to him as a young boy in Poland? He told me that he found himself moved by the military music, designed to reinforce patriotism. And as the Russians advanced and the radio every day would say only that the Germans were "straightening their lines," he knew this meant the war would soon be over, and he was glad for that.

So how many qualitatively different categories of consideration are at work here? (1) Dogs can be pets and "as such" are harmless. In a pack at night, something triggers their predator nature, and they be-come killers. (2) Food can be regarded primarily as a nurturing source of life and also used as a weapon when withheld. (3) A weed-free crop is regarded as desirable except when it stands in response to modified genes in the crop, which allow it to survive an industrial chemical, glyphosate, a probable cause of non-Hodgkin's lymphoma. (4) We are a product of the Enlightenment, made possible by a Roman empire lake surrounded by a diversity of cultures. (5) Lothar marched with and wore the uniform of the Nazi youth, was inspired by German mil-itary music, and was delighted that the Russians were close at hand and the war would soon be over.

So pet versus killer; food to nurture versus being used as a weapon; weed-free good versus poisonous field; cultural product of Europe due to an early Roman lake; Nazi youth member inspired by Nazi mil-itary music, happy to know the Russians would soon defeat what the uniforms and music stood for. How many stated conclusions might we now expect? Maybe confusion is a virtue that will make someone neither rich nor famous.

David Defeats Goliath

Now for one of the big questions always on my mind: How can a small number of people be most helpful in changing the course of history? At The Land Institute, both in Natural Systems Agriculture research and in the Ecosphere Studies education program, we wrestle with this, given what we believe to be the stakes in creating a sustainable agriculture and culture.

I learned a lot about that question from Marshall Ganz's book *Why David Sometimes Wins*.[5] His personal history is worth attention. He started as an undergrad at Harvard, dropped out, joined the civil rights movement, and then went to work for the United Farm Workers in California. Twenty-eight years after he dropped out, he returned to Harvard to finish a bachelor's degree in history and government, then a master's degree in public administration, and finally a PhD in sociology. With nearly three decades of field experience behind him and back in the company of scholars, he carefully analyzed various movement successes and failures. *Why David Sometimes Wins* reports on what he learned, both as an activist and as a student of movement politics.

His metaphorical question was something like this: How could a shepherd boy like David defeat the well-armed, seasoned giant warrior Goliath? In the Hebrew Bible (1 Samuel 17), we're told that Goliath has a large spear, a brass helmet, a coat of mail, brass on his legs. He cries out every day to the armies of Israel, "Choose you a man for your side. If he is able to kill me, we will be your servants, but if I kill him then shall ye be our servants." No one out of Israel's army steps forward. Finally David announces that he will take on Goliath. King Saul reminds David that he is a kid and Goliath "a man of war from his youth." So the king prepares David with his armor, a brass helmet on his head and a coat of mail, and girds David with his sword.

While David is getting dressed for battle, he must be thinking about the ill-fitting battle gear because once fully dressed, he tells

King Saul that he can't go into battle like this. He takes off all the hardware that Saul has lent him, goes to a brook, and selects five smooth stones, which he puts into his shepherd's bag. Now, with sling in hand and his bag of five stones, he approaches Goliath, reaches into his bag, takes out a stone, places it in his sling, slings it, and smites the Philistine giant in his forehead, whereupon "he fell upon his face to the earth."

There is more to the story of what happens next, but here is Ganz's analysis: David won because he had courage and because he did not think like Goliath. After accepting the equipment for battle, he took it all off because he had not mastered it. Instead, he developed a plan based on what he knew and had mastered.

So here is the sequence: courage first, followed by the ability to think differently about battle and to use equipment he had previously mastered. And here we must acknowledge that Goliath underestimates him. "Am I a dog, that you come at me with sticks? Come here," Goliath says, "and I'll give your flesh to the birds and the wild animals!" Plenty cocky, Goliath is.

At a small nonprofit in Kansas with only a handful of colleagues around the country, we seek to learn from that story. What follows is drawn from the first chapter of Ganz's book, titled "How David Beat Goliath."

Why is David, unlike everyone else on the battlefield, so strategically resourceful? He is more motivated, and he is angry that no one will respond to Goliath's insults. Ganz pointed out something very important for us: Once David is called to act, he "commits to the outcome before he knows how he will achieve it." That describes my approach to ecosphere studies—commit to changing the way people think about the planet and their place on it without knowing how that's going to happen. Ganz continued: David's "commitment to act does not depend on his knowledge of a feasible strategy. Rather he devises a feasible strategy based on his commitment to act. His decision to fight moves him to figure out how he can do so successfully." Commit, then figure it out.

In addition to learning from his experience, Ganz conducted a formal study of movements, reading and evaluating what other researchers had discovered. One of those insights: Motivation enhances creativity, and associated with creativity are concentration, enthusiasm, risk-taking, persistence, and learning.

Ganz put it this way: "When we are intensely interested in a problem, dissatisfied with the status quo, or experiencing a breach in our expectations, we think more critically and have a way to enhance our creativity, in part because they generate greater motivation." One can wonder why it takes extensive research to "prove" what seems intuitive. But it's nice to have our gut feelings validated. After all, our strongly felt and obvious conclusions can turn out to be wrong.

Another insight: The difference between extrinsic and intrinsic rewards is crucial. There are typically two kinds of extrinsic rewards: How much money will I get paid? How much fame will I achieve? But, according to Ganz, "the intrinsic rewards associated with doing work one loves to do, work one finds inherently meaningful, are far more motivating than extrinsic rewards."

For successful social movement leaders, their work is not a job but a "vocation" or a "calling." And what do you know, their rewards are intrinsic and highly motivating. Motivational differences can account in no small part for differences in resourcefulness among leadership teams. There is that word "resourceful" again, as we remember David's selection of the stones.

Here is where Ganz's experience comes in. With the United Farm Workers, Ganz worked with leaders who succeeded where the Teamsters and AFL-CIO had not. Ganz believed it had to do with "the depth of each team's collective commitment to the enterprise."

Back to David and the lessons about success from his run-in with Goliath:

- David knew how to use those stones in a slingshot, and to use them well.
- David's skill, his competence, freed him to consider novel applications, in this case, killing a well-armed giant.
- That skill required practice. (Can't you just see him as a kid out there day after day with Jesse's sheep, practicing with that sling on countless targets?)

What Ganz found in the scholarly literature was "that creativity in a craft is linked to mastery of its tools—that is, to the craftsperson's relevant knowledge and skill."

In reading all this, I could not help but think of The Land Institute scientists and their efforts to develop an ecological agriculture.

Their extrinsic rewards have not been great. Yes, they have achieved some status from their numerous articles and general attention the work has received. It has been endorsed by the National Academy of Sciences–National Research Council and the Royal Society of Great Britain. They have been invited to participate in Rome with scientists from nineteen other countries at the UN Food and Agriculture Organization meeting that was devoted to perennial grain research, all inspired by our work.

But what led up to the Rome meeting? Nothing less than a mastery of skills: Stan Cox's seed yields among the perennial sorghum hybrids, David Van Tassel's work with sunflowers and *Silphium*, Lee DeHaan's seed size and overall yield increases in Kernza®, Shuwen Wang's perennial wheat hybrids under evaluation in twenty sites in eight countries. And there's the work of Chinese colleagues we support to perennialize rice. And it's not just attendance at one meeting. Tim Crews, our ecologist and research director, presented our ideas to an Ecological Society of America meeting, which is some kind of a first, serving as an ambassador of sorts to an assembly not accustomed to thinking much about agriculture. An international meeting in Lund, Sweden, in 2019 included more than 90 researchers from all continents save Antarctica. No perennial grains there yet.

We believe we have a novel approach to the problem of agriculture in the same way that David had a novel approach to Goliath:

- We reimagined the field.
- We could have used conventional chemicals, the industrial approach, but it is not in us because we have an ecological worldview.
- As outsiders, we see ecological resources that others do not.

Industrial agriculture is a modern-day Goliath. Since we do not meet that Goliath on its terms, so far industrial agriculture does not see our work as a threat. When that changes, we'll face more tests of our commitment.

Living in the Industrial World

We have a love/hate relationship with the world we have made. We might as well get used to it.

Satan Is on the Other End

It seems that at one time, most in the Christian world believed that heaven was above and hell below. Dante's *The Divine Comedy*, which describes a field trip through hell, increased the imagination of some. Some imagined that it was dark down there, maybe in some places very black. No light at all. One could even imagine that Satan's official residence was below. What one believes has something to do with decisions made here on the surface. It seems that to stay out of that place called Hell requires certain beliefs and acceptable behavior.

And now my story begins. My wife, Joan, and I are going through Hillsboro, Kansas, a Mennonite community whose ancestors were the German-Russian immigrants said to have brought Turkey Red wheat to Kansas. We stop along the old highway on the west side of town to tour a house said to be somewhat like the houses they left behind in Russia.

We took that tour some twenty years ago, guided by a man long past his retirement years. Here is what we learned. Peter Paul Loewen and his family arrived at this spot about 1874 and by 1876 had this pioneer house built of air-dried adobe, with wooden beams on the inside. It has an elegant heating system—the cookstove fueled with prairie grass and wood, warm in the winter and cool in the summer.

I don't remember the guide saying how many generations and for how long others had lived in the structure. What is important here is that a certain Mennonite owner had rented this very adobe structure to the man who happened to be our guide. The guide told this story of how he and his new wife had set up housekeeping there, paying the landlord a yearly rent.

By the end of the first year or so, the young couple had their first child. A year or two later came the second child, which caused our guide with his young and growing family to ask the landlord if they might be able to have electricity, seeing as how with babies and laundry and farm work and such, electricity would be nice. This meant a washing machine, at a minimum, and any other necessary electrical appliance that comes to mind. They were happy with the elegant central heating stove, perfectly fine for warmth and cooking using wood, and the adobe house was relatively cool in the summer, but the family thought electricity would be nice. The owner said no.

The young renter asked the old Mennonite again if they might be allowed electricity. Still no. Somewhere along the way, trying to be persuasive, the young renter remarked to the owner how nice it would be to have electricity. The old Mennonite owner replied with something like "Ah, yes, it would be nice to have electricity, but Satan's on the other end."

This is obviously not a time to be too dismissive about what people believe and where their imaginations might take them. And so my opening about the depths of hell. It is surely a stretch, but I can't resist acknowledging that what was then fueling power plants was mostly coal, which is black. It comes from under the ground, ancient sunlight stored for multiple millions of years. The Industrial Revolution, which is now some 250 years old, began with water power, but fossil carbon as an energy source quickly followed and has been the mainstay from early on.

At this writing, it seems that power plants are the number-one source of greenhouse gases, which now stand at well over four hundred parts per million of carbon in our atmosphere. The consequence, with no end in sight, is what we call global warming, climate change, climate disruption, whatever. One may not like using Satan as a metaphor, but clearly the old Mennonite was way ahead of the rest of us. It is worth remembering that these are the same people who don't go to war, have come out against violence, help others in distress, and embrace community life. One doesn't have to endorse all their social practices or their theology to see some advantages in their tradition.

What if they had paid more attention to the rest of us and followed the modern path of "progress"? We would have lacked their good example. How appropriate was the metaphor, indeed? We are dependent on the dark energy that comes out of that dark place. Is the concept of Satan real or imagined in control of that dark stuff? Well, we moderns say, "We need more evidence."

It's not a tenet of Mennonite philosophy today to reject electricity, and it seems that the old Mennonite man did not claim to know what the consequences of Satan's approval of electricity would be. He just believed that Satan was "on the other end." Out of what scripture did he get the idea, or was it a direct message? I can't answer that.

These Mennonites had left tsarist Russia, and before that their kind had left other places in Europe. They were not followers of the Reformation ideas, and they had suffered under those folks as well. What was it about these Mennonites, and why did they come here? Well, we know the answer. It was the same for countless other immigrants—freedom of religion and of speech and land to own. Our Constitution and Bill of Rights allow large numbers of people to believe, if they want, that "Satan is on the other end" of an electric plug.

E. F. Schumacher Visits The Land, March 1977

Dr. E. F. Schumacher, Rhodes scholar and economist, was the author of a popular book titled *Small Is Beautiful*, published in 1973.[1] He toured America, promoting his book and his ideas. He paid a day-and-a-half visit to The Land Institute in March 1977 and died on a train the following August in Europe.

I found out that he was going to be in the United States from the president of Doane College in Crete, Nebraska, and invited Schumacher to visit and give a talk. Board member Sam Evans provided the Evans Grain Company plane for the trip from and to Nebraska. We had dinner with TLI board members in the home of founding board member John Simpson and his family and toured The Land.

In the evening lecture, Schumacher told some memorable stories. I will tell two of those in a moment, but first a bit about the walk around The Land. Mind you, our start-up had been established for scarcely seven months. On top of that, our classroom building had burned down six weeks after we started, with no insurance on the building, as the company was figuring out the relationship between our new nonprofit venture and the land our family owned. Having no money, we had been accumulating junk that seemed to hold promise someday for use in building this and that. We intended to have wind machines and solar collectors, which we eventually did. But here and there were iron and wood, fit mostly for most people's idea of a junk pile. We managed to acquire 225 patio doors from a company that had gone out of business. Some were double-paned, and some were single, but there they were in a pile as Schumacher and I toured the premises.

Excerpted and adapted from "A Story, as told by E. F. Schumacher," my contribution to the twenty-fifth anniversary edition of Schumacher's classic 1973 book, *Small Is Beautiful: Economics as if People Mattered—25 Years Later, with Commentaries* (Point Roberts, WA: Hartley and Marks Publishers, 1999).

E. F. Schumacher with the author during his visit to The Land Institute in March 1977.

The place looked worse than all of that, given that the terraces on half of the property had just been built, and here was this vast area with fresh dirt recently gouged out and thrown up. It did look terrible, and Schumacher asked with a tone of bewilderment, "My God, what is going on here?" I explained that the land had been seriously eroded, and I wanted to stop it. He quickly responded, "Have you thought of trees?" Well, I had read J. Russell Smith's book *Tree Crops*[2] and had fully intended to plant the Millwood variety of honey locust that reportedly would yield 1,800 pounds of sugar per acre.

But now, back to the patio doors. We paused in front of them, and I asked a question that had been bothering me for a long time: "Is it right, as we think of a sustainable sunshine future, to be drawing so heavily on the products of the Industrial Revolution since the Industrial Revolution is such a big part of the problem?" He said nothing at first, walked a few feet, then stopped, turned, and said, "Never mind; materials want to be used. They will show you how." Clearly, this was a response out of the Buddhist part of his mind. The book that had made him famous, *Small Is Beautiful*, featured Buddhist economics. He explained that it was about the same as Christian economics, but if he had called it Christian, "no one would want the book." Apparently Buddhism, at that moment, was a more intriguing label.

So right there was a contrast. The Buddhist had said that materials want to be used and will show you how, and my secularized Methodist Christian mind wanted to know whether we should be using this stuff that results from the fallen world we live in. So much for the tour. And now for his stories.

One of his most memorable stories had to do with lorries carrying biscuits from Edinburgh to London meeting lorries carrying biscuits from London to Edinburgh. "I wondered," he said, "that they don't exchange recipes." He added, "But then, I am just an economist, and there must be something about the added value of the transport." It was a great way of pointing out the incredible waste in capitalism.

I have saved his most memorable story for last. He told how he and some friends toured across the United States, right through our part of the country. It was at the height of the Great Depression, he said, when he and his companions came to a small town somewhere in Kansas and stopped at a service station. He engaged a local man there and asked him, "Well, how is it going?"

"All right," the local said.

"Well, what do you do?"

"I work on that farm over there," he said, pointing across the road.

"So you are a farmer."

"Yes, I used to own that farm, but I had no money to pay the hired hand, so I paid him in land. Now he owns the farm."

"Well, that is a very sad story."

"Oh, no, he has no money, either, and now he is paying me back in land."

Well, it was a charming talk. Our community theater had few empty seats that night. The locals in attendance that night got an early taste of what was to come out of The Land Institute. We still embrace Schumacher's values, making his remarks a perfect launch of what we already were and to a great degree did become.

The Shard and My Chevy Silverado

A few years ago, as I was driving east on Interstate 70 on my way to either Lawrence or Kansas City, I decided to detour a few miles north to drive by the farm where I grew up, something I do now and then. I don't usually stop, but this time I did and walked over the black soil characteristic of the Kansas River Valley.

It was forty acres, a small farm by today's standards, that at any one time had twenty to twenty-seven crops (my parents kept meticulous records), all irrigated when necessary from a well that yielded five hundred gallons a minute. From the pumphouse, the water was directed to the main ditch by a few ceramic tiles—short sections of pipe some thirty inches in diameter. The main ditch was connected to the fields by a series of branch ditches where we used our shovels to give the water advice, plugging here, opening there.

As I walked along on this particular late-fall day, long after harvest, I thought of my grandfather's purchase of the farm in the nineteenth century and the diversity of crops and livestock on that land when I was growing up. Those subjects are never far from my mind on that land.

The barn is now gone, along with the chickens, geese, hogs, horses, milk cows. Also gone are the woods on an adjacent property, the trees along the creek, the trees of a windbreak on which fifty or so turkeys would roost. Alongside that windbreak we planted "hotbeds," a method for planting early, when it was still cold, to make sure that the young plants didn't freeze. Glass, elevated off the ground by boards or pipes on the sides, was laid over the seedlings to trap the heat.

An aside: Much later, thinking about those cobbled-together hotbeds, I looked online at pictures for hotbed design. I shouldn't have been surprised to see that today's gardeners can buy hotbed boxes with slick wood frames and hinged glass tops that are far fancier than the ones we built on the farm with whatever materials we had avail-

Strawberry field on the Jackson family farm being irrigated in the 1930s. Farmstead buildings appear in the background, with the school more distant on the left.

able. I also learned that the term "hotbed" is used almost exclusively for a technique using considerable amounts of manure—as it decomposes, the manure generates heat to warm the plants. I searched my memory but couldn't remember us ever using manure in our hotbeds. It appears that the meaning of the term as we used it had changed over time, now describing not only the method of covering but the source of heat.

Back to that black Kaw Valley soil. While walking the land that day, I spotted a shard of that ceramic tile. To me, that was a more exciting find than unearthing a prehistoric archaeological treasure. This shard has no value to anyone but me, but it hangs in my office as a favorite art piece. I love it, I suppose as a placeholder. About the memory it conjured up, I can't say. But the shard and I have an

My brother Elmer at age seventeen, irrigating the corn fields in 1936.

asymmetrical relationship. The shard doesn't return my love. In the same way, the earth may not seem to love us, at least on our terms. But isn't it enough for us to love it in all of its little pieces?

The subject of unreciprocated affection makes me think of an old-when-I-bought-it Chevy Silverado half-ton pickup with four-wheel drive.

It's not hyperbole to say that I loved that truck. It accumulated many miles and many dents from work around The Land from our early days on. I loved it, though in some ways I was not kind to it. After years of abuse due to the countless jobs our work required—hauling, pushing, pulling, driving anywhere no matter what the land surface and sometimes through water or whatever needed to be dealt with—the day came when it was no longer able to do what needed done.

The engine was shot. The most recent set of tires was thin. I had welded the rear bumper a few times. It was unpredictable how long second gear would stay put. There was no way to get it into shape to sell to anyone else. It was ready for the junkyard, where I finally drove it and collected the standard $50 from the office. Then I walked around to the back for one last look, just in time to watch the jaws of the crane capture it from the sides, squeeze it in at the doors, break the glass, lift it over the other mountains of scrap metal, and drop it from what was, I thought, an unnecessary height.

Sad does not begin to capture what I felt. The equipment operator, I imagined with little thought, probably felt nothing as he moved the metal around. Even so, I wondered, why could he not have waited until I left? I felt strong affection for what had served me, my family, and The Land Institute so often and so well. But that affection was all one way, from me to it. I am reasonably sure that truck cared not a whit for me.

Wendell Berry, borrowing from E. M. Forster, reminded us in his Jefferson Lecture that "it all turns on affection."[3] It is considerably easier to care for the people and things we love than to care for what we are indifferent to. And it is also easier to maintain affection and commitment when they are reciprocated. We have affection for each other and for our places, and we now have to think about our relationship to the earth as a whole, even if it doesn't seem to care about us.

As the old hymn "O Worship the King" goes, the king is "our Maker, Defender, Redeemer, and Friend." Most would think of the king as Christ and the Lord. But can we not recognize that the earth is our maker? After all, it is from what we call the "nonliving" earth that life emerged. That's the journey from minerals to cells, and from those cells came, through Darwinian selection, the vast diversity of life—all part of "the journey of the universe." That helps reorient us. If we don't want to speak of Creation (by a divine force), we can speak of the earth as our Creator. The earth, literally, made us, and it continues to defend us with an atmosphere that protects us. Whatever story we tell reminds us that the patterns of this world emerge from the earth, not from our heads. I repeat: We humans do not make the patterns of the world.

The earth is our maker and defender, and now with proper attention we can participate in our redemption—the restorative, regenerative work of renewal. But that requires us first to recognize that no one can be redeemed before acknowledging failure, before confronting our own sin, whether born of innocence or evil.

Is the earth our friend? Does the earth care about us? That depends on our understanding of friendship. Do we want the earth to be the sycophantic friend who always tells us we are smart, good-looking, and funny? That won't help much. We need the earth to be the kind of friend who tells us the hard truths about ourselves. If we listen to the earth for that friendship, perhaps it will be there.

Thoughts on the Natural History of Eden

In the late 1960s, while teaching at Kansas Wesleyan University, I would drive around Saline County looking at rural property for a small homestead, hoping to round out what I considered the perfect life of teaching at a small liberal arts college, coaching track, and raising the three children my wife and I had planned and brought into the world.

On several occasions, I found myself parked a few miles south of town by an old iron bridge over the Smoky Hill River. On the east side was a high bank that overlooked the river and a beautiful floodplain opposite. The strip along the high bank most attracted me. It had once been broken and farmed but had since returned to native grassland. Two ravines, one major and one minor, cut their way toward the river, and from them spread trees such as burr oak, green ash, black walnut, hackberry, and box elder. There were the usual accompanying poison ivy and grapevines. Gray dogwood and sumac spilled into the prairie.

It was an idyllic spot. My favorite place was a high point where I could look down on a ripple created by an outcropping of Wellington shale. To sit there was an exquisite experience. I am tempted to say that I meditated on the wonders of nature, but I doubt that. I don't know what I thought. I do know that I was always alone. I never felt like having anyone with me. I was not interested in hunting the land's pheasant, quail, cottontails or squirrels. Nor had I any desire to fish the stream. The place was Eden.

I learned that an older, childless couple, Bessie and Loyd Wauhob, owned this little strip. They lived across the river and across the road. I went to see them and expressed my interest in purchasing a small piece nearest the road, maybe three acres. Bessie's dad had told her

Excerpted and adapted from Wes Jackson, *Consulting the Genius of the Place: An Ecological Approach to a New Agriculture* (Berkeley, CA: Counterpoint Press, 2010).

never to sell any of her farmland, and she had stuck to that. While both Bessie and Loyd agreed with my assessment that the land was so erodible that it had to be abandoned as crop ground, they remained reluctant to sell. They were not the kind of people one should push. But I returned to visit them a few times and offered as much as $1,000 an acre for three acres. They finally agreed to sell but protested that $1,000 an acre was too much and that I need pay only $500. A deal was struck.

Of course, I wanted to start building a house and raising the children in the country right away. My wife agreed. The permit office asked to see my plans. Using paper on the counter, I drew the entire perimeter freehand and added a couple of doors. I declared, "Here are my plans." They were approved.

For several months, we did nothing to the place, lacking the money to start. My family and I frequently came out to picnic or walk around "The Land." (That is what we called it and how The Land Institute got its name.) We decided to put the house perpendicular to the setting sun on the winter solstice and more or less parallel to a slope toward the river north of the ravine, from which poor-quality coal had been mined more than a half century before. I built a construction shack out of damaged dormitory doors. I added a $25 lab bench from Kansas Wesleyan's old science building. Water came from a well that my geologist friend Nick Fent drilled across the road. Power and water lines went to the shack in the same trench.

The International Harvester dealer rented me an industrial tractor with a backhoe and a front-end loader at $5 per tractor hour. I went to work digging the basement, which was to be a walkout affair, filled at the back but mostly open on three sides. After more or less mastering the backhoe, I dug the ditch for the lateral field and the hole for the septic tank. I dug the footing for the house—two to three feet wide and about that deep. Miscellaneous pieces of iron, including old bicycles and tricycles, reinforced the concrete walls poured mostly into self-built forms, intended to be ten inches thick. (I measured one recently, and it is closer to eleven inches.) The unconventional construction included local trees for beams and inside panels.

For rest, I often ambled over to the river, to the same spot where I had stood before construction began, and looked down and out and around. Nothing below had changed. The Wellington shale still generated the ripple. The rustic iron bridge, which should never have been a part of Eden but somehow was, still spanned the river. The fields across the water were the same. To my left remained the large woody ravine. But Eden was gone. I tried to bring it back by opening and shutting my eyes, to imagine what it was, but it never returned or even came close. Apparently, the very exercise of what makes us human, in that place, drove what some would call "the spirit" away from me. A philosopher might call it a phenomenological experience. I thought of the biblical meaning of the angel with the flaming sword who denies access to Eden.

Art Zajonc's book *Catching the Light*[4] helped me understand that the very design of the experiment in trying to determine whether light is a wave or a particle determines whether one will perceive a wave or a particle. The design and the perception are products of our cognition. There is a drawing, an optical illusion, that sometimes appears as a young woman and at other times as an old woman. You can't see both at once. In the case of wave versus particle, maiden or grandmother, it goes back and forth. For me, and my place of Eden, a cognitive switch had been thrown that has never been thrown back.

One interpretation of Genesis may be that our fallen condition comes from insisting that we participate in the Creation. Because I participated in the Creation as a *technological creature,* I had destroyed something whole, which is to say holy. My family and I, like all others, wanted a home. Who can argue against that motivation? Had I met the shelter need in a minimal sort of way, would Eden have remained? We'll never know. I do know that my perceived need at the time was probably not a real need. We could have continued to live in town. My perceived need was determined more by culture and desire to live in the country than by necessity.

Bess and Loyd continued to sell us adjacent land as we could afford it until we eventually owned twenty-eight acres. Back from the river now grow various trees bearing organic fruit—beautiful cherries and pears and also wormy apples. The deer and wild turkey have increased since the early days. The habitat is still safe for them as well as bobcats, quail, pheasants, and nonverbal serpents, not at all tempting us and less onerous and certainly less toxic than the poison ivy along the big ravine. Nearly five decades have passed since I first experienced Eden there and lost it. Now near the river bluff is a merry-go-round for grandchildren. A sweat-of-the-brow flower and vegetable garden grows over the lateral field, and the shade and pleasing forms of native and exotic trees planted as saplings early on—all watered from that well Nick Fent drilled—bring delight to this participant in the Creation who sometimes thinks the loss of Eden was a bargain.

What Is to Become of Us?

What do we need to be if there is to be a future?
Who do we need to be?

The Day I Discovered that I Am a Groupie

I'm fond of pointing out that we are a species out of context, meaning the patterns of the world we live in are not the patterns of the world in which our species evolved. The genus *Homo* is about 2.5 million years old, and the species *Homo sapiens* is about 200,000 years old. Before the domestication of plants and animals began 10,000 to 12,000 years ago, human beings were largely gatherers and hunters. After agriculture, everything changed except us. We're still essentially the same animal, just living in a dramatically different world. Everything around us—gadgets, buildings, highways as well as forms of political organization, religions, races, and classes—is new. We are a species trying to live in a world we built, but it's a world that in many ways was not built for the kind of animals we are.

One not-so-obvious point: For virtually all of our evolutionary history, we humans lived in small band-level societies, typically in the range of fifteen to fifty people. Those societies did not have hierarchies as we know them today. There were no kings and queens bossing people around, no homecoming kings and queens to be jealous of. More on that in the next chapter.

One obvious point: For virtually all of our evolutionary history, we humans moved across the landscape under our own power. We walked and ran.

That is why I'm a devoted fan of track and field and why I love to spend time with others equally devoted. I'm honoring my species' evolutionary history. That's all important to the story of one amazing footrace. First, a bit of backstory.

The University of Kansas has a reputation for recruiting and turning out exceptional track and field athletes. The late Bill Easton was controversial, but he knew how to recruit and coach. The KU relays each spring bring in top competitors year after year. During my high school days, I was a great fan of various top performers. The Glenn

Cunningham Mile was named after a great Kansas runner whose 1934 world-record time (4:06.8) stood for three years. In 1936, Glenn Cunningham also set a world record of 1:49.7 in the 800 meters, overcoming a lot of adversity along the way. When he was in grade school, his legs were severely burned in a schoolhouse fire that killed his brother. Doctors told him he would never walk again. In addition to being one of the world's premier runners, he went on to earn a PhD from New York University in 1938. A great Kansan.

In the mid-1950s, there was much interest in the possibility of running the mile under four minutes. Kansas had a fast young man from ranch country in the western part of the state near the town of Ashland, Wes Santee, who was called the Ashland Antelope. He was considered America's leading candidate to be the first to break the four-minute barrier, but British doctoral student Roger Bannister got there first in 1954. Kansas soon had a first, though. As a senior at Wichita East High School, Jim Ryun became the first high school student to break the four-minute barrier.

Billy Mills was a member of the Oglala Sioux Tribe, part of the Lakota Nation, from the Pine Ridge Reservation in South Dakota, where he had been orphaned at age twelve. He had gone to the Haskell Institute, now Haskell Indian Nations University, in Lawrence, Kansas. After his days at Haskell, Mills trained under Easton and eventually joined the marines, and when the Olympics were held in Tokyo in 1964, he qualified for the 10,000-meter run. Billy was relatively unknown before the event and was not expected to finish in the top three.

But as the race progressed, Mills was keeping up with two other runners ahead of him. One summary of the event put it this way: "Then suddenly, as if an apparition had come upon them, Billy Mills sprinted past both and won by three metres." His victory took everyone by surprise. "Asked if he was concerned about Mills, the third-place finisher replied, 'Concerned about him, I never heard of him.' One Japanese reporter asked Mills, 'Who are you?'"[1]

Were someone to ask me about the most exciting race ever run, ruling out the meets when I coached both high school and college track, it would have to be this race for Mills, who won in what is widely considered one of the greatest upsets in Olympic history.

Tom Rupp, who had been a distance runner at KU, later coached the track and cross-country teams at Sacred Heart High School, which was across the street from the Kansas Wesleyan University track and field facilities. Tom and I worked together in a wonderfully cooperative manner during those few short years when I was teaching and coaching track at Wesleyan.

Tom eventually quit teaching and coaching and went into private business in Salina. One Sunday morning, up early and in my easy chair reading the *Salina Journal*, I saw that Tom was hosting former KU track and field alumni right here in Salina for the entire weekend. Here is where I learned that I was a groupie. I called Tom and told him I had been reading the paper about those former great stars. He immediately said they were all at his house at that moment and were about to have breakfast. "Come on over, Wes." I went.

Well, it was a great morning talking about all the great moments in track history at KU and what everyone was doing now.

I told Wes Santee that I had wrecked my five-tone-green Dodge coupe in Lawrence and spun a '51 Mercury around in the intersection of Ninth and Mississippi—all to watch him run on what turned out to be a wet track, even though he had a "rabbit" for the first three-quarters. (The "rabbit" was a fellow runner who kept the pace for three-quarters for Santee to follow and then fell back and left Santee to finish the last quarter-mile sprint on his own.) Well, he failed to finish under four minutes, making it a disappointing moment, for sure.

But now about Billy Mills. After breakfast, he and I sat alone in the kitchen, and I told him that his victory in Tokyo represented a huge emotional high for me and how I wept without shame as he sprinted down the stretch. The movie *Running Brave* is about Mills's life and that race. During the few times I have seen the film about his challenges, I've found it impossible not to tear up. Mills was modest. His reply to my praise, my admiration, my recounting moments within the race and his strong sprint across the finish line, was simply this: "Well, Wes, it wasn't me."

This was a sacred moment for me, and I did not ask, "What do you mean?" or "How so?" I wasn't sure what he meant, but it didn't seem like my place to probe further. Instead, I excused myself and went to the restroom. What could I say, remembering how he had set the world record that day for the 10K race at 24 minutes, 24.4 seconds?

I've met a lot of very smart people in my life, including Nobel Prize winners. I've met talented artists. But my short conversation with Billy Mills is one of my most precious moments. Call me a groupie if you like, and I suppose it's accurate. But I also think there's something else at work in this story.

We are all more creatures of the Upper Paleolithic than we are of the Internet age. No matter how much time any one of us spends behind a desk or the wheel of a car, we really were born to run. How much of the activity at a track meet is simply modern people yearning for a return to their evolutionary context?

And what of Mills's humility? Why was I so moved by it and so unwilling to ask him to analyze it? Why was I so happy to be included in that small gathering of track and field folks? Was it, again, simply my desire to be in the kind of relationship to a fellow human that was the norm in that same evolutionary context?

Am I just trying to divert attention from the more parsimonious explanation, that I'm just a track and field groupie? Perhaps, but I'll press my point a bit further. At least in the affluent sectors of the developed world, we've lived with the assumption that there's always more. But I believe the future is going to be defined by living within limits, with learning to adapt to less. In other words, our success in the future is not in returning to the past but in getting closer to being a species in context. One part of that process involves looking at our current lives for the echoes of the Upper Paleolithic.

We can look back to how those ancestors managed pride and jealousy. We can look at how outliers today have escaped the siren call of consumerism. More about that in the next story.

The Necessity of Insulting the Meat:
Ferocious Egalitarianism

Back to the question of jealousy and resentment. The world we live in is full of hierarchies. Bosses and workers, rulers and ruled. Billie Holiday sang it in "God Bless the Child": "Them that's got shall get / Them that's not shall lose / So the Bible said and it still is news." Hierarchy everywhere, so much so that we can forget that it really is new, an outgrowth of what I have called the "problem *of* agriculture."

Most of human history has been lived in small band-level societies of gatherers and hunters, where hierarchy is not the norm. What can we learn from that?

What I am about to describe comes from a 2017 *New Yorker* article by John Lanchester[2] that drew on books by James Scott (*Against the Grain*)[3] and James Suzman (*Affluence without Abundance: The Disappearing World of the Bushmen*).[4]

My takeaway from that reading was the importance of encouraging a "ferocious egalitarianism" in our efforts to create a sustainable society.

Suzman reported from his research in Africa that the most valuable thing a hunter can do is come back with meat. This is not the case with gathered plants, since they are "not subject to any strict conventions of sharing." According to Suzman, for the bushmen of southern Africa (also known as the San people), hunted meat is carefully distributed according to protocol. The tribal members who eat the meat given to them have an "insulting the meat" ritual. They make it a matter of practice to be rude. Most of us would consider that offensive and inappropriate. If someone offers a gift, why would you take the gift but insult it?

But the practice makes perfect sense in a society that rejects hierarchy. Tribal members don't want the hunter to get too uppity. To prevent the idea of considering yourself better than anyone else, the

group has to promote that ferocious egalitarianism. They are out
to block or eliminate any path to tribal disruption. Suzman wrote,
"'When a young man kills much meat,' a Bushman told anthropolo-
gist Richard B. Lee, 'he comes to think of himself as a chief or a big
man, and he thinks of the rest of us as his servants or inferiors. . . . We
can't accept this.' The insults are designed to 'cool his heart and make
him gentle.'" For these hunter-gatherers, Suzman said, "the sum of
individual self-interest and the jealousy that policed it was a fiercely
egalitarian society where profitable exchange, hierarchy, and significant material inequality were not tolerated."

Here is the final paragraph of the *New Yorker* article:

> This egalitarian impulse, Suzman suggests, is central to the hunter-
> gatherer's ability to live a life that is, on its own terms, affluent, but
> without abundance, without excess, and without competitive acquisi-
> tion. The secret ingredient seems to be the positive harnessing of the
> general human impulse to envy. As he says, "If this kind of egalitari-
> anism is a precondition for us to embrace a post-labor world, then I
> suspect it may prove a very hard nut to crack."

"Hard nut to crack," "old chestnuts," "hardheaded realism." We've
all heard these utterances of mature judgment, common sense, or a
simple "Good luck on that one." Most often they are uttered to stop
the conversation or end thought.

Well, the fact is that members of our own species came up with
"fierce egalitarianism." It wasn't an idea that descended from the heav-
ens. It's an example of a practice that recognizes the way in which
an individual weakness can harm group solidarity. In that context, it
seems to work. Our context is different—no argument there. Time to
go to work finding examples that can work in our world.

Leland

> We are vastly superior to any other species in stretching our
> world into the shape we want; that also makes us infinitely
> more capable of creating unforeseen difficulties. As a general
> rule, the greater the changes we think into being, the more
> problems we have to face. Environmental history is, among
> other things, a lengthy account of human beings over and
> over imagining their way into a serious pickle.
> —*Elliot West*, The Contested Plains

At the species level, *Homo sapiens* has demonstrated no sign that we
have the ability to practice meaningful restraint. Like bacteria on a
petri dish, fruit flies in a flask, or rabbits without predators, if the energy is there, our numbers increase. As powerful creators of abstrac-
tions, we have developed an economic system (capitalism) that, as
it stands, moves our population and the spending of materials and
energy as fast as possible to the edge of the petri dish. Having been effective at death control through diet and medicine, we are now forced
to exercise restraint. Where do we turn?

We have a few examples of people whose lives serve as examples
of alternative ways of living, and we can imagine one day a Chamber
of Resilience replacing the Chamber of Commerce, trading in our
obsession with growth for a sense of responsibility.

My friend Leland Lorenzen died on September 6, 2005, not far
from his seventy-ninth birthday. He had lived on less than $500 a
year for nearly all of the twenty-nine years that he lived in a shack,
six by sixteen feet. A small wood-burning stove provided heat. He ate
mostly soaked wheat, greens from his yard, and from time to time

Excerpted and adapted from Wes Jackson, *Consulting the Genius of the Place: An Ecological Approach to a New Agriculture* (Berkeley, CA: Counterpoint Press, 2010).

Leland Lorenzen outside of his shack. Photo by Terry Evans.

milk from his goat. He died as he had lived. A day or two before he died, he turned to his son, Jule, and said, "Time to open a hole." The family buried him in his sleeping bag on his one-acre plot, his grave dug with a fossil-fuel-powered backhoe by a neighbor who refused to accept pay for the digging.

It has been my experience that those who come the closest to "walking the talk" don't actually talk about it. Leland was the best example I know of a "walker," but he, too, was dependent on the extractive economy, including for his burial. Even Leland was never completely off the grid. He acknowledged that he was grid dependent, especially when he was a merchant seaman. During those seven years at sea, he had been around the world in various ports, which contributed to his radical views on economics. In ports near and far he noticed the difference between the economy of the street and the economy of the official culture.

Once settled and working at the local oil refinery, he read Thoreau's *Walden*. What he had been turning over in his mind came together. He tossed a copy of the book on the kitchen table and told Bernice, his wife, that she had better read it because it was going to change the life of their family, which included three children, at the time ages four, eleven, and thirteen. During this period, he was quite apocalyptic, certain that nuclear war was inevitable.

Beyond *Walden*, one of Leland's insights was that we start doing violence to people and the environment when we seek pleasure. He insisted that there is nothing wrong with the experience of pleasure, but when we seek it, we start manipulating the world, people included. Conflicts between individuals or nations come from the same source, he often said. Here is how it works. In our heads, we have an imagined environment that will bring us pleasure, and through the pursuit of pleasure, we begin to try to duplicate it in the world, which others have to live in with their own imagined environment. This leads to conflict. The pursuit of making the imagined environment real, and of that pleasure, is in conflict with the imagined environment and pleasure other people are pursuing. In such a manner, we destroy the world's fabric.

To most people, I suppose Leland carried his beliefs to an extreme. When I first knew him, he had a beautiful garden and, in the winter, a hotbed of sorts that extended the growing season of various vegetables. Then one year he quit gardening, reckoning that this, too, was a form of pleasure seeking. From then on it was greens out of the yard, soaked wheat, and milk from his goat.

He shortened a rusted-out Volkswagen Karmann Ghia convertible and drove it to conserve fuel. It looked like Donald Duck's car. During his tenure in the shack, he figured he spent $350 on gasoline and tobacco and $150 on food each year.

Leland lived thirty miles south of The Land Institute, relatively isolated in the country. Late one winter, on my way to the Wichita airport to catch a plane, I left enough time to stop by his shack to see how he was and to visit. I noticed that he had been poking seeds into a flat of dirt. I asked Leland, "What's going on here? You're going to have a garden again. You're back into pleasure seeking."

"Come on in, Wes," he said. "I'll tell you all about it. I'm all [screwed] up."

According to Leland, Bernice, who lived in a bomb-shelter-type house some seventy feet away, felt that she needed $300 a month to meet expenses. Her Social Security check was $200, but because Leland had worked at regular jobs before reading Thoreau, he was eligible for $400 a month. So Leland took out Social Security and was giving Bernice $100 per month. The remaining $300 was "piling up in the bank." That accumulation was causing him to have "creative thoughts," and he had started various projects in his shop and on his place. Indeed, these thoughts had led him into pleasure seeking again. His brain was "on fire with imagination." He began to imagine pleasure foods that he knew he could do without.

I couldn't help him out of his dilemma, and besides, I had to get to the airport and contribute to more atmospheric carbon. This was March. Leland continued to draw on his Social Security until around Christmas. My daughter Laura had come home for the holidays, and on a cold Christmas Day, she and I drove toward Leland's. We planned to visit the Maxwell Game Preserve to check on the herd of bison and prairie elk, have a winter prairie picnic there, and then go see Leland. We parked our car at the preserve and walked a couple of miles through pasture, fields, and snow to Leland's. He was in his shack, and once again he began to complain about the money piling up in the bank forcing him into creative thinking. Laura said, "Well, Leland, why don't you just quit taking that Social Security since it is bringing you so many problems?" A month or so later, Leland drove up to tell me, "Laura's words kept ringing in my ears, and I'm going to scratch my name off the Social Security list."

He told me later it wasn't as easy as he thought. He tried, but the official told him that if he did, he would have to give back all that he had received. Well, of course Bernice had spent her allotment, and Leland had spent some on his "creative efforts." Leland said he couldn't give it back. The official told him to give it to a worthy cause. "There are no worthy causes," he said. "Give it to your children," the official suggested. "They aren't worthy either," he replied. (I think he thought it would be a source of problems for them because I know he loved them dearly.) He was stuck. The money had become a curse.

This idea of money as a curse reminds me of another incident involving Leland more than a decade earlier. I was putting a roof on what became the new classroom at The Land Institute after the other had burned down. Leland stopped by and, crawling up the ladder, began to help. It was inexpensive roll roofing. He helped me immensely that day, and I had both a twenty-dollar bill and a five in my billfold. I tried to give Leland the twenty. He refused. I then tried to give him the five. He said no. I insisted, saying that I could afford it and that he was to take it. The conversation ended when he said, "Don't give me your problems."

Countless numbers of people have asked me what he did with his time. I do know that he had an intellectual life. He would go to the public library, check out six books, read them in the shack, return them, and get six more. He had read a lot of the Indian philosopher Krishnamurti and knew the Bible pretty well. Though he had serious doubts about an afterlife, he liked a lot of what Jesus had to say and had a particular fascination with the Hebrew Scriptures. He once said to me, "When you have Jews come around The Land Institute, I want to meet them if they take their religion seriously, especially the Orthodox Jews."

A major effort for Leland was to stop the internal dialogue. "We are always either building or protecting an image," he told me. He also thought it nearly impossible to rid the mind's desire to do so in the presence of another person. This was a source of worry. He said the only way out of it was to be alone; after a while, his image of himself would fade, and then he would have the "awareness of a squirrel." A squirrel's awareness is of the "effective," immediate environment surrounding him. One can then be out of the environment where the buds of violence would grow.

He took me once a couple of miles from his shack to an abandoned pasture with prairie and trees all around where there were some large protruding rocks. Using those rocks and a minimum of building materials, Leland had built a small shelter just large enough to sleep in. It was really spare. The pillowcase was stuffed with prairie grasses. Here the deer and other wild animals would come to lie down outside, within a few feet of where he was sitting or sleeping, undisturbed by his presence. As I surveyed the surroundings and inspected his handiwork, he explained, "Here is where Leland goes to get away from Leland." I didn't ask what he meant and still think it would have been improper to do so, but I have wondered. My first question was

"Why isn't life back at his shack enough?" I thought of Francis of Assisi, some of the mystics, some monks, Elijah, and other examples. Perhaps even when alone, an image of who we are begins to form.

Once when I brought a "certifiable intellectual" to visit with Leland, the visitor was disappointed. He said I had given Leland too much credit and thought that Leland had nothing to offer. I didn't argue. It was good enough for me that Leland had become one of my indispensable friends.

He once told me that the wheat he ate daily cost him about three cents and that when he had the goat for his milk, the wheat that went into that goat cost nine cents. The value of the wheat came to less than $45 per year. Now and then he would mix honey into his soaked wheat. He called honey a pleasure food that he could do without. His method of fixing the wheat began by measuring out what he would consume the next day. He would soak it in water and cook it with a forty-watt lightbulb for a few hours. (Bernice's house has electricity.) The cover for his container was a hubcap.

Once we took a driving trip in a pickup truck to Long Island, New York, to bring home tools and equipment that had been donated to The Land Institute by a friend whose husband had died. On this trip, Leland declared that he wasn't going to let me eat him "under the table." But I did. He got so sick that we had to stop at a grocery store and buy him some peanut butter and crackers. Back to a simpler diet, he was fine. On another trip one January, he took me around to visit some dropout communities in the Missouri Ozarks. Most of the people had been antiwar protesters, civil rights activists, and the like who had thrown themselves out on the land, taking their advanced degrees from places such as the University of California, Berkeley with them. It was near-subsistence living. It was a good trip, and I enjoyed the conversation. Leland and I talked about that trip many times, noting how difficult it is to try to become a satellite of sustainability orbiting the extractive economy. Over the years, most of these idealistic, strong individuals found themselves increasingly pulled into the orbit of the dominant culture.

Once Leland told me that his moments of depression, which sometimes lasted a few days, came as a gift of sorts. He said that it was like a "great cleansing," that it was followed by great clarity and insight. Like Thoreau, he had many visitors who were attracted by his philosophies. Several months after his death, The Land Institute purchased the shack from Bernice, moved it thirty miles north to The Land Institute, and refurbished it a bit. It now stands as a monument to the most bottom-line person I have ever known.

Visitors today can see taped to the wall of Leland's shack a government form. He had initialed each "I DO NOT WANT: surgery, heart-lung resuscitation (CPR), antibiotics, dialysis, mechanical ventilator, tube feedings." Then he wrote, "or any other health care treatment. I wish to heal or die naturally. Please take me to my bed." *Please* is underlined three times, and *my* is circled. He had two witnesses.

Ten years and three months later, he died. He was buried the same day on his one acre. After that neighbor with the backhoe had dug the grave and Leland was laid to rest, the family covered the grave themselves. It did not cost the family one penny.

... make this Health Care Treatment Directive to
...... my health care and to provide clear and convincing
...hen I lack the capacity to make or communicate my decisions)
...that I will regain such capacity.
...a certain life prolonging procedure or other health care treatment
...relieve pain or lead to a significant recovery, I direct my physician to
...nable period of time. However, if such treatment proves to be ineffec-
...be withdrawn even if so doing may shorten my life.
...health care treatment to relieve pain or to provide comfort even if such treat
...en my life, suppress my appetite or my breathing, or be habit-forming.

...e prolonging procedures be withheld or withdrawn when there is no hope of
...covery, and I have:
...al condition; or
...tion, disease or injury without reasonable expectation that I will regain an accept-
...e quality of life; or
bstantial brain damage or brain disease which cannot be significantly reversed.

When any of the above conditions exist (I DO NOT WANT) the life prolonging procedures
which I have initialed below. (You should assume any treatments not initialed may be administered to you.)

- surgery.. JLL initials
- heart-lung resuscitation (CPR).................................. JLL initials
- antibiotics ...;... JLL initials
- dialysis... JLL initials
- mechanical ventilator (respirator)............................. JLL initials
- tube feedings (food and water delivered through a tube in the vein,
  nose or stomach... JLL initials
- other   Or any other Health Care Treatment .... JLL initials

2.) I make other instructions as follows: (You may describe what a minimally acceptable
quality of life is for you.)
   I wish to heal or die naturally. Please take me
   to (my) bed.

te: _____   X Signature J. Leland Lorenzen

tness _Gerald E. Payne_   Date 7/30/95   Witness _____ Date 1 May 95

tarization

Notarization of the Durable Power of Attorney is required in some states (e.g., Missouri but not Kansas). If this
document is both witnessed and notarized, it is more likely to be honored in other states.

this ___1___ day of ___May___, 199_5_, before me personally appeared the aforesaid declarant, to me
own to be the person described in and who executed the foregoing instrument and acknowledged that he/she
ecuted the same as his/her free act and deed. IN WITNESS WHEREOF, I have hereunto set my hand and affixed my
icial seal in the County of __Dickinson__, State of __Kansas__, the day and year first above written.

Cheryl Stieben   3-10-99
ry Public   My Commission Expires

ceptance: (Optional) I have discussed this document with the person making this durable power of attorney and I
ept the responsibility designated to me as stated above.   CHERYL STIEBEN
                                                                  State of Kansas

The government form where Leland requested he be allowed to die naturally in
bed. Photo by Terry Evans.

Conclusion: Hardening Off

I once planted some seeds of a wild winter annual in small pots in a greenhouse. They were painstakingly watered and fertilized and produced a green, luxurious growth, surpassing in overall vegetative vigor their relatives in the field.

From experience, we knew that if we moved these plants from the cozy greenhouse environment and left them outside, they would be vulnerable to the very environment that had shaped their ancestors. A high percentage would be unable to withstand the shock and might die, not because they lacked the genetic potential to resist the environmental extremes but because the narrow greenhouse environment had not called forth the broad spectrum of genetic potential necessary to endure the adversity usually presented to wild populations.

The United States as a developed country might be regarded as a greenhouse culture. Lately we have been watching a gathering storm outside our comfortable environment and have become suddenly cognizant of how vulnerable our culture is. We have reason to believe that our cozy environment may fast disappear. In fact, only a few supporting subsystems responsible for our affluence need falter, and we will find ourselves "out in the cold."

This essay appeared in the first issue of *The Land Report* in December 1976, https://landinstitute.org/wp-content/uploads/2018/06/LR-1.pdf.

Brothers Dwight and Elmer, with author standing. All five of my siblings (two sisters who were the oldest, and three brothers) were far more "hardened off" than I, having come through the Depression when I was just an infant. Their cultural capacity was better aligned with meeting needs long after the Depression with their minds at work. Harley's judgment of the professed need for a new European Union building (see "Brother Harley at the European Parliament") was one example of the interest and range of my siblings' engagement as it pertained to Need and "How to Do." Courtesy of David Plowden.

There is a way to gradually prepare greenhouse plants for a full life outside. It is called "hardening off." If the plants are placed outside for a few hours each day at the beginning, and the amount of time they are left outside is gradually increased, eventually they can be safely left outside permanently. The first time they are placed outside, on a quiet, warm afternoon, the outside environment may appear to differ very little from the greenhouse environment. But it is an important first step, and somehow it seems different. What we are doing during this hardening-off period is giving the plant the outside conditions and the time for its genetic machinery to kick in and enable it to cope physically and psychologically with the outside environment.

One might say that our main purpose at The Land Institute is to provide alternatives to the present for a cultural hardening-off process. Student work on projects described on these pages may be no more than moving plants from a warm, still greenhouse to the outside on a warm, still day. But we hope most of our activity is a bit more than that. We know that if we jump too quickly into the world of the future, we might become so discouraged that we will refuse to venture out again. We hope that one day we may regard being whipped by the wind as being touched by the earth rather than threatened with wilt, but that can happen only if we have been properly hardened off.

Notes

Foreword

1. Mary Oliver, *Red Bird* (Boston: Beacon Press, 2008), 37.

2. Jonathan Gottschall, *The Storytelling Animal: How Stories Make Us Human* (New York: Houghton Mifflin, 2012), xiv.

3. Ian Tattersall, *Masters of the Planet: The Search for Our Human Origins* (New York: Palgrave Macmillan, 2012), xiv.

4. Muriel Rukeyser, "The Speed of Darkness," in *The Speed of Darkness* (New York: Random House, 1968), stanza 9, lines 3–4.

Introduction

1. Robert Lee and Tristan Ahtone, "Land-Grab Universities: Expropriated Indigenous Land Is the Foundation of the Land-Grant University System," *High Country News*, March 30, 2020, https://www.hcn.org/issues/52.4/indigenous -affairs-education-land-grab-universities.

One Thing Leads to Another

1. William Faulkner, *Requiem for a Nun* (New York: Vintage Reprint, 2012), 73.

2. Morris Berman, *Dark Ages America: The Final Phase of Empire* (New York: W. W. Norton, 2011), 236.

3. John Wesley, *The Journal of John Wesley* (Grand Rapids, MI: Christian Classics Ethereal Library), 55, https://www.ccel.org/w/wesley/journal/cache/jour nal.pdf.

4. John L. Heatwole, *The Burning: Sheridan in the Shenandoah Valley* (Charlottesville, VA: Rockbridge Publishing, 1998), 8.

5. US War Department, *The War of the Rebellion: A Compilation of the Official Records of the Union and Confederate Armies* (Washington, DC: US Government Printing Office, 1902), 822.

My Rural Life

1. Wes Jackson, "Letters of a Humble Radical," in *Wendell Berry: Life and Work*, ed. Jason Peters (Lexington: University Press of Kentucky, 2007), 165.

2. Wendell Berry, *The Gift of Good Land: Further Essays, Cultural and Agricultural* (New York: North Point Press), 82.

Schooling, Formal and Informal

1. Irving Stone, *Lust for Life*, 50th anniversary ed. (New York: Plume, 1984).

2. "The Origins of Glassmaking," Corning Museum of Glass, https://www.cmog.org/article/origins-glassmaking; "Glass Timeline," History of Glass, http://www.historyofglass.com/glass-history/glass-timeline/.

3. Liz Logan, "How Pyrex Reinvented Glass for a New Age," *Smithsonian* magazine, June 5, 2015, https://www.smithsonianmag.com/innovation/how-pyrex-reinvented-glass-new-age-180955513/.

Scientifically Speaking

1. Hans Jenny, *The Soil Resource: Origin and Behavior* (New York: Springer-Verlag, 1980).

2. Arnold Schultz, "Ecosystemology: Systems Thinking for the Earth's Future." https://nature.berkeley.edu/sites/default/files/Arnold%20Schultz%2C%20Ecosystemology%2C%202009.pdf.

3. Wes Jackson and Wendell Berry, "A 50-Year Farm Bill," *New York Times*, January 4, 2009, https://www.nytimes.com/2009/01/05/opinion/05berry.html.

4. Hans Jenny, *Factors of Soil Formation: A System of Quantitative Pedology* (New York: McGraw-Hill, 1941).

5. Theodosius Dobzhansky, *Genetics and the Origin of Species* (New York: Columbia University Press, 1937).

6. N. V. Tsitsin, "Remote Hybridization as a Method of Creating New Species and Varieties of Plants," *Euphytica* 14 (1965): 326–330.

7. Sarah Mae Brown, "Russian Workers Cope as Best They Can," *Salina Journal*, March 13, 1997.

8. Harold J. Morowitz, *Cosmic Joy and Local Pain: Musings of a Mystic Scientist* (New York: Scribner, 1987).

9. Steven Weinberg, *The First Three Minutes: A Modern View of the Origin of the Universe* (New York: Basic Books, 1977), 154.

Ideas I've Run into along the Way

1. Richard Levins and Richard Lewontin, *The Dialectical Biologist* (Cambridge, MA: Harvard University Press, 1985).

2. Adam Gopnik, "Inquiring Minds: The Spanish Inquisition Revisited," *New Yorker*, January 16, 2012, https://www.newyorker.com/magazine/2012/01/16/inquiring-minds.

3. Joseph Swift, "The Tie That Binds," *Science* 355, no. 6326 (February 17, 2017): 70, https://science.sciencemag.org/content/355/6326/701.full.

4. Peter Watson, *Convergence: The Idea at the Heart of Science* (New York: Simon and Schuster, 2017).

5. Marshall Ganz, *Why David Sometimes Wins: Leadership, Organization, and Strategy in the California Farm Worker Movement* (New York: Oxford University Press, 2009).

Living in the Industrial World

1. E. F. Schumacher, *Small Is Beautiful: Economics as if People Mattered* (New York: Harper and Row, 1973).

2. J. Russell Smith, *Tree Crops: A Permanent Agriculture* (New York: Harcourt, Brace and Co., 1929).

3. Wendell Berry, *It All Turns on Affection: The Jefferson Lecture and Other Essays* (Berkeley, CA: Counterpoint, 2012).

4. Arthur Zajonc, *Catching the Light: The Entwined History of Light and Mind* (New York: Bantam Books, 1993).

What Is to Become of Us?

1. These quotations come from an article, "Athletics at the 1964 Tokyo Summer Games: Men's 10,000 Metres," that is no longer online (https://www.sports-reference.com/olympics/summer/1964/ATH/mens-10000-metres.html). A similar account is Earl Gustkey, "Mills' Miracle: When He Won the 10,000 at Tokyo 30 Years Ago Today, It Might Have Been the Greatest Upset of All Time,"

Los Angeles Times, October 14, 1994, https://www.sports-reference.com/olympics/summer/1964/ATH/mens-10000-metres.html.

2. John Lanchester, "The Case against Civilization: Did Our Hunter-Gatherer Ancestors Have It Better?," *New Yorker,* September 11, 2017, https://www.newyorker.com/magazine/2017/09/18/the-case-against-civilization.

3. James C. Scott, *Against the Grain: A Deep History of the Earliest States* (New Haven, CT: Yale University Press, 2017).

4. James Suzman, *Affluence without Abundance: The Disappearing World of the Bushmen* (New York: Bloomsbury Publishing, 2017).